0 to Physics in 60,000 words

Trevor Boardman

Copyright © 2019 Trevor Boardman

All rights reserved.

ISBN: 9781702769075

DEDICATION

For Sarah

CONTENTS

	Acknowledgements	i
	Introduction	1
1	Energy	3
2	Charge in Solids	22
3	Light part 1	48
4	Cosmology part 1	66
5	Astrology	78
6	Atom part 1	80
7	Luminescence	94
8	Charge in Fluids	98
9	Light part 2	106
10	Atom part 2	113
11	Chemistry	143
12	Nucleus	155
13	Atom part 3	172
14	Cosmology part 2	188
	Epilogue	204
	References	207
	Glossary	211
	Index	215

ACKNOWLEDGEMENTS

My interest in all things scientific began when my older brothers gave me a box of electrical parts and wires and showed me how they could be fitted together to make things happen. I am indebted to the staff of Leeds University Physics department who inspired me and showed me how the theories fitted together.

I need to thank Mary Boardman and Kate Lambe for pointing out what is likely to be understood by the layman, Matt Boardman for text improvement, and Mark Boardman and Alice Evans for illustration advice. Finally, I need to thank Mary again for undertaking the arduous task of proofreading my text.

INTRODUCTION

Homer Simpson once told Lisa "In this house we obey the laws of thermodynamics!" He was right and that goes for all the other laws of science in all houses throughout the universe since its beginning and until its end no matter how alien you think you are. Any alien society with an interest in what is in the universe and why it is the way it is and with technology as advanced or more advanced than ours (if they exist) will probably have discovered science in the same way we did. It would certainly be the same science.

This fact that science applies to the whole universe for all time and that it explains nearly all of why the universe is the way it is, is what many people find compelling about the subject.

This book is aimed at people who are interested in why the universe is the way it is, or want answers to questions like "Why do they think dark matter exists?", "Why doesn't electricity leak out of electrical sockets?" "How do they know that galaxies rotate?", "Why are some things transparent", "How do they know the universe is 13 billion years old?".

The answers to these questions involve different layers of scientific knowledge which can be daunting at first, but by starting with more simple concepts and building up from there you will quickly get the hang of it. We'll also look at the fascinating history of the concepts: the people and their stories that led to scientific breakthroughs.

All of the science that we'll look at in this book is part of physics, which underlies all other areas of science. Physics itself is based on maths, but in this book we're going to avoid mathematical explanations in favour of qualitative descriptions, which are often easier to understand.

Physics has several threads running through it which developed alongside each other and provide insights for each other. This book has been structured around them with one or more chapters to each thread. It begins with the first observations mankind wrote down about science and builds knowledge in the sequence it was discovered. It has a structure which should allow the reader to jump in at any point if that is what is wanted.

Before you begin, a few words on science itself. Science can be thought of as a jigsaw puzzle, each discovery and insight is one of the pieces. Some progress was made by finding new pieces and some by fitting pieces together. In finding and fitting those pieces scientists often act as detectives looking at clues to find the culprit. This jigsaw puzzle reveals a beautiful picture which explains more and more of things in the universe that are too big and too small to imagine. This includes things that engineers can use to make our world more comfortable and fun.

As yet, no straight edge pieces of the puzzle have been found. Perhaps they never will.

So on to the physics.

1 ENERGY

Energy, it flows through wires, through space and every cell in our bodies needs it to keep us alive. It is something people spend 11% of their incomes on in the form of gas and electricity. But what is it? For everyday life all we need to know is that we buy some and it makes our cars go, keeps our house warm and lights our streets. But for science, gaining a deep understanding of energy has been fundamental to all progress.

However, energy is not an easy topic. The concept was developed by the German philosopher Gottfried Leibniz in the 2nd half of the 17th century and 100 years elapsed before it was generally accepted.

So, what is it doing at the start of a book that is supposed to make physics easy to understand? The main thing about energy, as far as science is concerned, is that it underpins everything else. If you can understand energy you can probably understand all of science. So, this is the deep end. But fear not dear reader. We are not going to jump. Our pool is nice and warm, has some easy steps and we are going to take those.

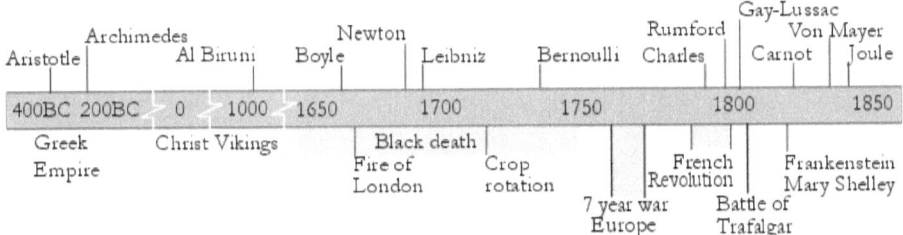

What is a force?

Before we start talking about energy it is necessary to say a few words about force.

It is an idea that was introduced by ancient Greek philosophers Aristotle and Archimedes. They said that a force was necessary to start a stationary object moving. This fits with our experience. When you slide a pepper pot across the table you apply a force and it stops as soon as you stop pushing.

This is how things stood until the 17th century when British mathematician Isaac Newton realised this was not the whole story. For example, when you throw a stone it keeps on going after it has left your hand. He extended the Greek ideas with his 3 laws of motion.

The idea of a scientific law sounds scary but don't be alarmed. All it means is that scientists have looked at a thing many, many times in different circumstances and seen that the same thing always happens. The claim is then made that this will be true everywhere and at any time in the past or the future. One might make the observation that if you hit a ball with a bat the ball will go flying off. This has always been true and always will be true in all places. We could call it the bat and ball law. Every time a machine operates hundreds of these laws are effectively retested. If ever a law were to fail, planes would fall out of the sky, cars would stop and life would cease. But don't worry, this is not going to happen.

The first of Newton's laws of motion said that an object will keep moving at the same speed until a force acts on it to change its speed[1]. This effect we know as inertia.

---

[1] A force can also change the direction of an object

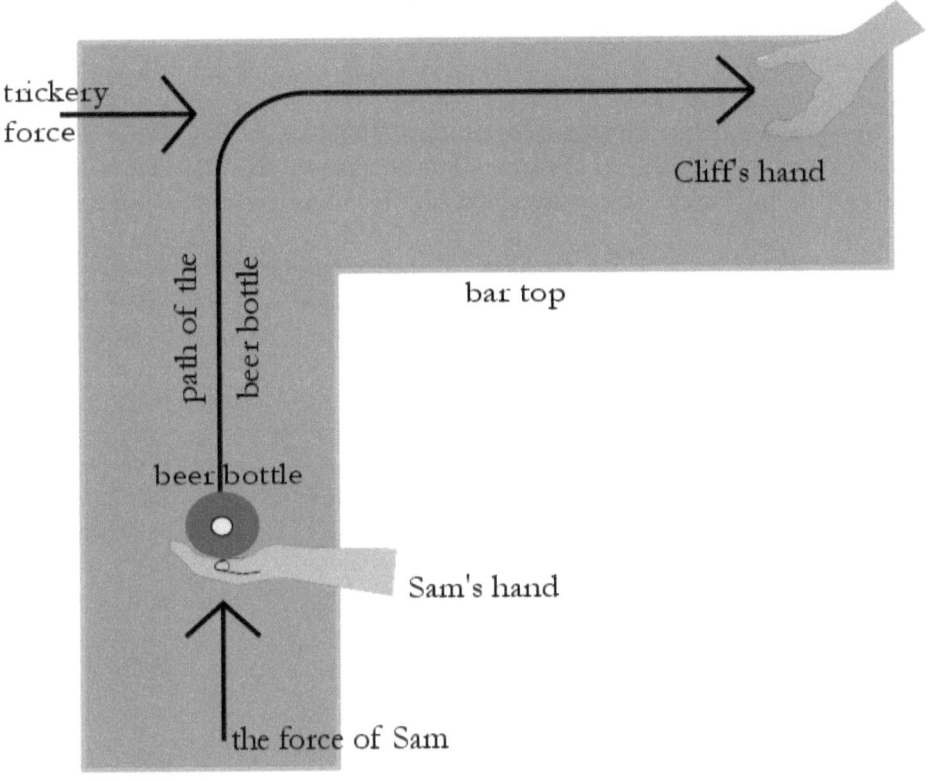

Fig 1-1 Cheers bar slide – Newton's 1st and 2nd laws of motion

But that is not quite what we observe in our lives. When Sam did his beer bottle slide in Cheers the bottle was slowing down all the time after it left Sam's hand and just about stopped by the time it reached Cliff's hand. This is because of friction which we will examine later in this chapter. However, if you could apply a force to a beer bottle in outer space where there is no air or countertop to cause friction the simple situation holds true. The bottle will keep on moving forever (assuming nothing gets in its way) or until another force acts on it to stop it. Another aspect of the beer bottle slide was that the bottle went round the corner of the bar to reach Cliff. For that to happen some clever behind the scenes trickery had to be employed which applied a sideways force when the bottle reached the corner. This highlights another aspect of Newton's 1st law which in its complete form says a moving object will continue at the same speed *and direction* after being set in motion. To change the direction of a moving object you have to apply a force from the side.

A daily example of a force changing the direction of an object is when a car goes round a bend, the front wheels are moved to point in the direction you want to go. This causes some of the force pushing the car forward to be diverted into pushing from the side. Another way to make a car go round a bend would be to directly push it from the side, but this is far from convenient.

If Sam had wanted to send a larger bottle along the bar, he would have needed to push it harder. Newton's 2nd law of motion allows people to calculate just how hard they would need to push. Or as scientists say, how big the force needs to be to get the desired acceleration.

Refining that idea a little, we would say the bigger the mass you want to move the bigger the force you will need. Also, the faster you want the mass to move the bigger the force you will need. If you want to make a heavy thing move fast you will need a really big force.

The 3rd of Newton's laws of motion says that whenever a force acts, an equal one acts in the opposite direction. Now this sounds suspicious. It is not at all what you would expect.

One way to look at this is that things inside the 'forced' object feel the opposite force. When your car accelerates you feel a force pushing you back into your seat. When the car corners left you get thrown to the right.

Another way to look at the 3rd law is to imagine two equal sized rocks in outer space. They are close together with a stick of dynamite between them. The dynamite explodes, and the two rocks fly off in opposite directions.

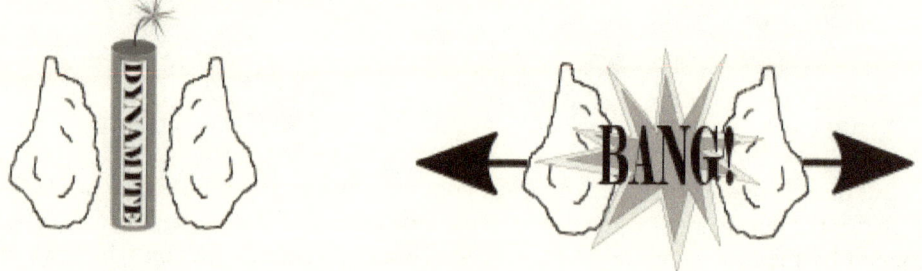

Fig 1-2 Equal mass rocks flying with dynamite exploding between them

Do the same thing with two rocks of different sizes and the same thing happens, except, due to Newton's 2nd law of motion, the small one flies off with more speed than the big one. Both rocks experience the same force, but the smaller rock has less mass, so it accelerates faster.

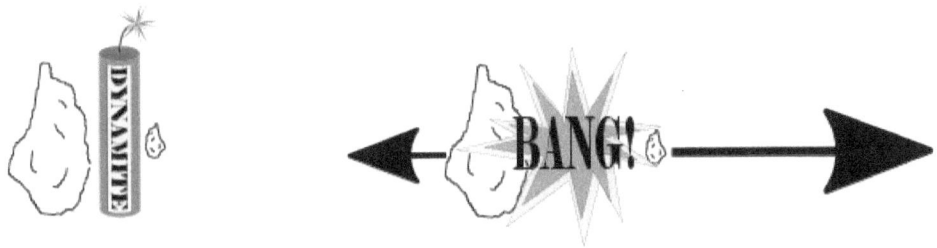

Fig 1-3 Unequal rocks with dynamite exploding between them – Newton's 3rd law of motion

Now replace the stick of dynamite with a spaceman. If he throws the small rock his arm is acting like the stick of dynamite. The small rock flies off fast but the spaceman plus large rock go off in the opposite direction.

Fig 1-4 Spaceman throwing rock – Newton's 3rd law of motion

Bringing this down to earth, if you throw a rock off the back of a boat, you and the boat move forwards. If you punch a wall your fist will hurt because the opposite force has acted on your fist. In all these cases, when a force acted the opposite force also acted. Newton's 3rd law says this is true every time a force acts in every situation.

When you swing a bucket half full of water around, the water stays in the bottom and you encounter the 1st and 3rd of Newton's laws of motion. The 1st tells us the water wants to travel in a straight line, so you are constantly changing is direction and consequently constantly pulling on the bucket handle. The 3rd law says there will be an equal and opposite force and that's what keeps the water in the bucket. This outward force is centrifugal force.

These laws of motion made it possible for the first time to make reliable predictions about the future, such as where a cannon ball is going to land. Today Newton's laws of motion are still used in all mechanical engineering and when plotting the trajectory of a spacecraft.

Additionally, Newton had an insight about what was going on when an apple fell off a tree. Previously, the only forces known were when people threw things or the wind blew but Newton realised that when an apple fell it did so because of a hitherto unsuspected force, the force of gravity which pulls apples from trees towards the earth. Maybe his thoughts went along the lines of "if I could see an apple fall in Brazil, which is about a quarter of the way round the earth, it would fall sideways". And "if I could see an apple fall on the opposite side of the earth it would appear to fall upwards to me". So, he got the idea that apples fall because there is a force attracting apples and everything else to the planet earth.

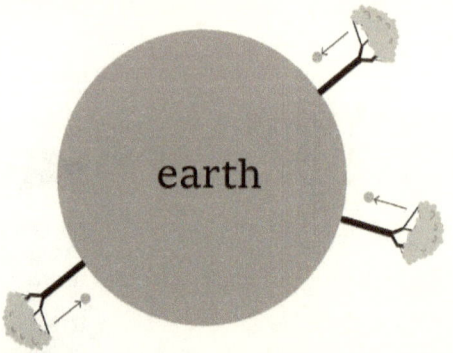

Also, crucially, the force of gravity is not the sole preserve of planet earth. All things everywhere attract stuff with the force of gravity.

Two objects in outer space like a spaceship and an asteroid will attract each other. The same thing happens on earth. All the things in a room are attracted to each other by gravity, but, because the earth has so much mass it pulls so hard on the objects in your room that the attraction between objects is not noticed. The force of gravity between two objects in a room was first measured by Henry Cavendish as will be seen in *Chapter 14 – Cosmology part 2*.

Another thing Newton realised is that the gravitational force is bigger for objects that are close together and for objects that are more massive. From this he created an equation which allows people to calculate the force of gravity between any two objects if you know their mass and the distance between them.

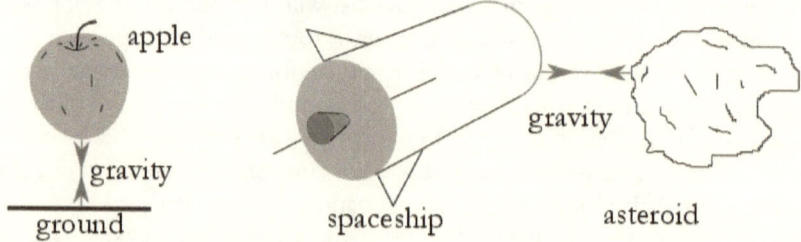

Fig 1-5 Objects attracting each other via the force of gravity

While Newton was formulating his laws, German mathematician Johannes Kepler was examining a large record of planet positions with their dates and times which had been compiled by the Dane Tycho Brahe. Kepler discovered big patterns in this data and from them he established a set of laws that described how planets circle the sun (more of this in *Chapter 4 - Cosmology part 1*). Some years later Newton and Leibniz independently developed a new type of maths (calculus) which allowed people to study the rate of change of things. This calculus enabled Newton to explain Kepler's laws in terms of his laws of gravity and motion. In doing so he provided further evidence that his laws of gravity and motion work the same way between objects in space as they do between objects in your room and on earth. The gravity that pulls the earth to the sun works in the same way as the gravity pulling an apple to the ground. The centrifugal force that keeps water in the bottom of a bucket when it is being swung around works in the same way as the centrifugal force that keeps the earth from falling into the sun. When a planet is in orbit around the sun the force of gravity matches the outward centrifugal force.

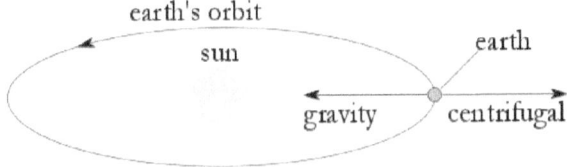

This is true wherever one object orbits another object: moons orbiting planets; planets orbiting stars; stars orbiting the centre of a galaxy (*Chapter 4 - Cosmology part 2*).

Probably the biggest visible manifestation of gravity is the ocean tides. The first person to associate the tides with the moon was the Greek astronomer Seleucus who recorded, in 150 BC, his observation that tides coincide with the location of the moon.

Newton's law of gravity readily explains the ocean bulge which gives us a high tide when the moon is overhead. But there are actually two bulges in the ocean, one on the side of the earth facing towards the moon and another on the side facing away from the moon. This second bulge mystified people until Newton brought his laws of motion to bear on it. We are familiar with the idea that the earth rotates around the sun once a year and it spins on its axis once every 24 hours. But there is a third motion of the earth. It wobbles once every 29 days and that wobble is caused by the moon. So why is that?

You can only balance a pencil on your finger if your finger is at the middle of the pencil. If you stick a lump of plasticine to one end the point at which the pencil balances moves towards the plasticine. The middle of the pencil above where it balances is known as the centre of mass.

Fig 1-6 How the centre of mass of a pencil moves if plasticine is stuck to one end

Imagine an axle with a wheel on each end made from parts of a construction toy. If you were to balance it on your finger the centre of mass would be in the middle of the axle. If you were to spin the axle round it would rotate around the same place. Now if you put some plasticine on one wheel to make it heavier the centre of mass would move towards that wheel and that would be the place where it balances and also the point about which it spins.

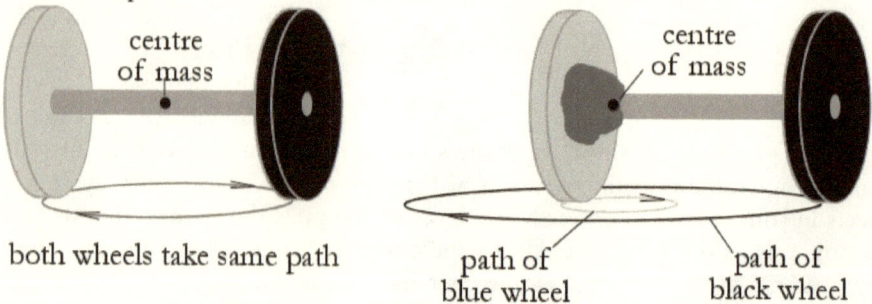

Fig 1-7 Two wheels on an axle will wobble if its centre of mass is moved to one end

This is similar to what happens when a moon orbits a planet. If the moon and planet have the same mass, they will orbit about a point midway between the two. If the planet is heavier the centre of mass of the pair of them will be closer to the heavier one. In the case of the earth and our moon the centre of mass is inside the earth. So, the earth and moon rotate around this point and the earth wobbles as the moon goes round it. A hammer thrower wobbles in the same way as he swings the hammer round before letting go.

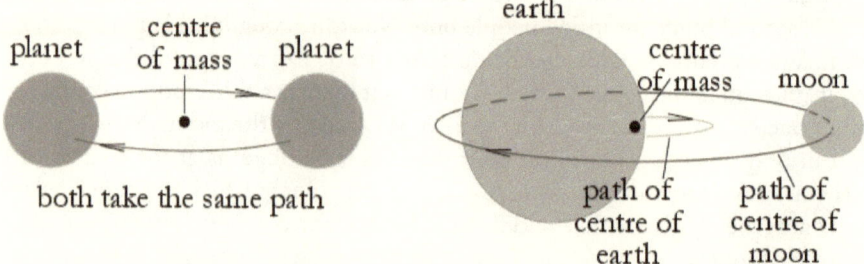

Fig 1-8 Objects orbit around their centre of mass so if one is much bigger that will wobble

The earth's wobble imposes a centrifugal force on the ocean on the side of the earth away from the moon and thus the second bulge occurs. This is another occurrence of Newton's 3rd law of motion (every force has an equal and opposite force). The force of gravity causing one tide is balanced by the equal and opposite centrifugal force causing the ocean to bulge on the side of the earth facing away from the moon.

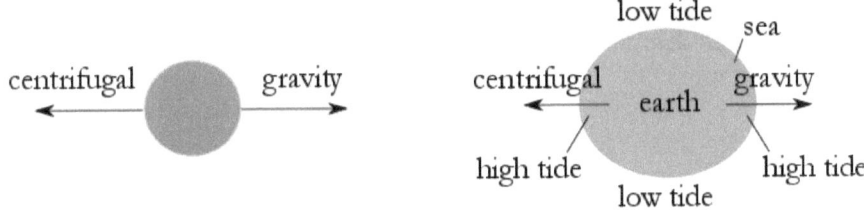
Fig 1-9 Earth's wobble creates a centrifugal force equal and opposite to the gravity of the moon

Newton's laws of motion showed that whenever something moves or changes speed or direction in our universe it does so because of a force. He also knew there were other forces besides gravity: forces that cause wind; forces that allow people to throw things; electric force; magnetic force. And today we would add the force that comes out of a fuel tank to make our cars go. What Newton didn't know was that all those other forces are just one force in disguise as will be seen in later chapters.

But where do these forces come from? The answer to this is bound up with the concept of energy.

What is energy?

Energy, Leibniz realised, is what is needed to make a force happen. The more energy you have the bigger the force you can have and the further you can act on a thing with that force.

You may be thinking, hang on, when I buy energy from my utility company most of it goes into keeping my house at a comfortable temperature. What have forces got to do with that? The connection between forces and heat will be outlined later in this chapter. But before we do that, we need to have a look at what this energy stuff is.

Imagine someone parks a car at the base of a cliff. The driver gets out, climbs the cliff and kicks a rock off which falls down and dents the car roof. Most people would concern themselves with the motivation for this act and the ownership of the car but here this little story will be used to explain energy and its relationship to force.

One way to picture energy is that it is the ability to do damage. If you drop a rock onto the roof of a car from a height of 1cm it probably won't even damage the paintwork. If you were to kick it off the top of a cliff so it dropped 10m on to the car it might make a dent say 10cm deep. The rock exerted a force on the car while the 10cm dent formed. After that all of the kinetic energy in the rock appears to have been expended. Anything that is moving can hit another thing and dent it or make it move. In doing so it has exerted a force. We can say an object has energy because it is moving. If anything gets in its way or it reaches the thing that is attracting it, a force will be acted on it. The amount of energy is the distance it could exert the force for. The rock dropped from 1cm had very little energy. The rock which fell 10m had enough energy to exert a force big enough to dent the car and to keep that force going for 10 cm.

And what happens next? We tend then to think that the energy has been used up after it has done its stuff like denting a car, but as we will see below, the real picture of what happened to that energy is quite different. Energy does not get used up but there is more than one type of energy and energy can change from one type to another.

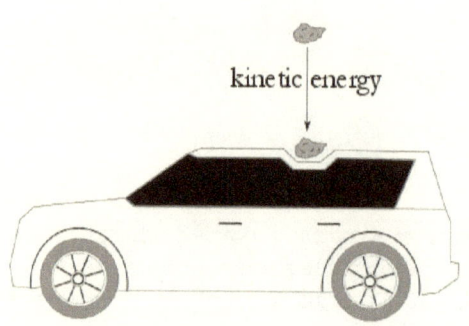

The energy a moving thing has is known as kinetic energy. Something that has kinetic energy can exert a force if it hits something. The more mass and/or speed it has the more kinetic energy it has.

When the rock was 10m above the car it had no kinetic energy but it was in a position where if it went off the cliff it would gain kinetic energy as it was attracted to earth. When a thing is in a place where it can feel a force and it has the potential to move under influence of a force, scientists say it has potential energy.

So, there are two distinct kinds of energy: kinetic energy (stuff that is moving) and potential energy (stuff that is in a place where something is pulling it).

When you buy energy, like a tank full of petrol, you buy it in the form of potential energy and when you use it you convert it to kinetic energy in some machine like a car.

An object that has potential energy has that energy when it is in a place where it has the potential to move because it is feeling some force. But it stays still because some other opposite force is stopping it.

The rock at the top of the cliff stayed still because the countless particles that make up the cliff are providing the force which is stopping the rock from going down to the centre of the earth under the force of gravity. The force provided by those particles can be 'removed' by kicking the rock off the cliff. Now, in mid-air, the rock no longer feels the force of the particles in the cliff and down it goes. The potential energy is now being converted to kinetic energy.

Another situation where an object has potential energy is an archer pulling an arrow in his bow. The arrow is stationary because the force in his arm matches the force in the opposite direction caused by the bending bow and its string. When the archer lets go the arrow takes flight. Or, if someone cuts the string, his arm flies back and the arrow drops to the ground.

Potential energy exists wherever there are two equal forces pulling an object in opposite directions so that it stays still. If one of those forces is removed the object goes flying off. This fits well with the archer's arm and the bow. But also, when the 1kg rock was on top of the cliff, the cliff was pushing up on it with a force of 1kg while gravity was pulling down on it with a force of 1kg, so it stayed still.

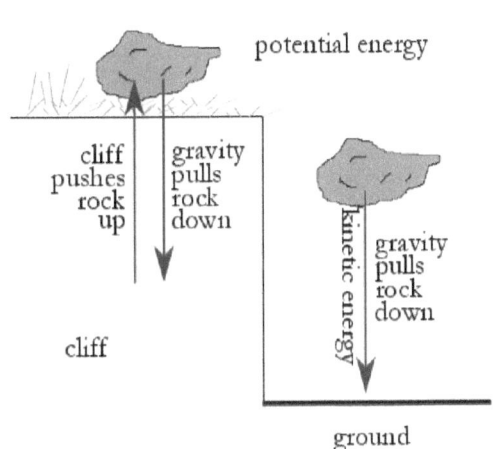

So how come the ground pushes up with a force of 1kg? How does it know it has to do that? It all sounds suspicious.

One explanation is that it is predicted by Isaac Newton's 3rd law of motion. The earth and rock in the cliff is a bit springy. Everything is a bit springy, even a steel girder. When you push down on a spring it compresses. As it compresses it pushes back. The harder you push the more it compresses and the more it pushes back. This is the way bathroom scales work. When you step on them you compress a spring. The more you compress the spring the more it pushes back. This goes on until the force it pushes back with is the same as the gravity that causes you to press down. That is, your weight. The dial shows how much the spring has compressed. The cliff pushes back with the same force as the mass of the rock (1kg) once it has been compressed a little bit. As Newton's 3rd law says, every action has an equal and opposite reaction.

In *Chapter 13 - Atom part 3* we will see why this springiness happens and why everything is springy.

The amount of potential energy an object has is equal to the kinetic energy it would have when it reaches the thing that is attracting it. The stronger the force and/or the longer the distance to the thing that is attracting it the more potential energy it has.

In the case of the rock held 10m above the car it has potential energy equal to the kinetic energy it would have when it hits the car.

In the case of the arrow it has potential energy equal to the kinetic energy it would have when it leaves the bow.

We have seen that turning potential energy into kinetic energy is as easy as falling off a log. Can it go the other way? Yes, but it is not quite so easy to arrange. Basically, you have to arrange for the kinetic energy to move something to a place where two forces will act on it in opposite directions. You could hit the rock with a big bat to get it from the ground to the top of the cliff. That is easy to imagine but difficult to arrange. It is easier to carry it up.

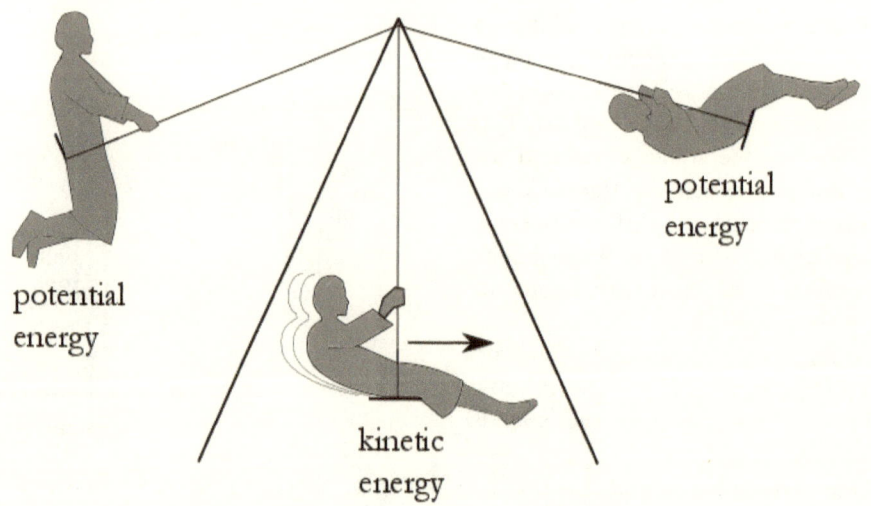

Fig 1-10 Person on a swing

An example of kinetic energy being converted to potential is a person on a swing. When she is furthest out, she is higher up and has potential energy. As she swings in and speeds up this potential energy gets converted to kinetic energy. At her lowest point she is moving fastest and has the least potential energy and the most kinetic energy. As her swing continues, she slows down as she gains height again. When she reaches the other end of her swing she stops. All her kinetic energy has been turned into potential energy again. She had been moved into a position where the force of gravity can act on her again.

So much for kinetic and potential energy but this does not explain what forces have to do with keeping your house warm. To explain this, we need to look at heat or, as scientists would say, heat energy. As we will see below, heat energy is in the kinetic energy camp.

What is heat energy?

In the 11th century Persian scholar Abu Al Biruni was the first to document that things can be made warm by rubbing them together. Friction impedes things' progress and increases their temperature. Normally we seek to get rid of friction in our machines so they will go faster, except when we are driving along at a nice speed and we come up against a line of stationary traffic. Then we are happy for the brakes of the car to turn its kinetic energy into heat energy. This is what friction does.

Several attempts were made to explain what heat is by saying it was a kind of fluid that flowed through stuff. That idea was used at first to explain electricity too and it didn't work there either. But in 1738 Swiss mathematician Daniel Bernoulli developed his kinetic theory of gases which works very well at predicting how a gas behaves (it gets hotter or colder and its pressure goes up or down when you compress or expand it). The core idea of this theory is that in a gas, molecules are darting about in all directions and obey Newton's laws of motion. These molecules will be explored in *Chapter 6 – Atom part 1*. One of the things Bernoulli's theory tells us is that temperature is a measure of the average speed of the molecules in a gas. The difference between hot things and cold things is that the molecules of hot things are moving about faster. In a gas they fly about all over the place, in a liquid they vibrate and slowly move amongst each other and in a solid they vibrate about a fixed point. In *Chapter 13 - Atom part 3* we will see why these three, so called, phases of matter exist.

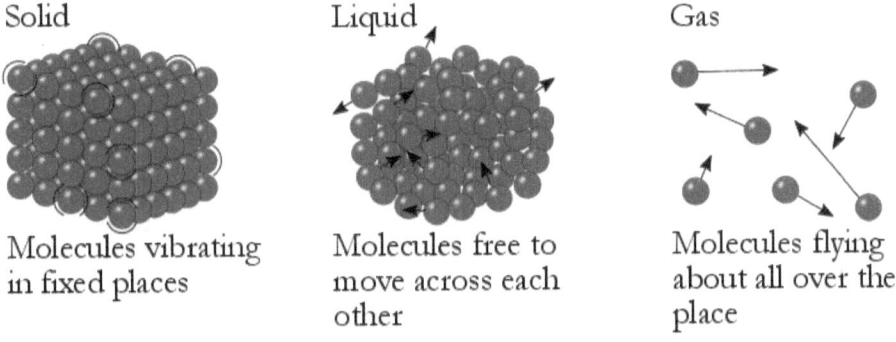

Fig 1-11 Three phases of matter

Over the years up to 1802 French scientists Jacques Charles, Joseph Gay-Lussac and Edme Mariotte and Irish scientist Robert Boyle made observations about how the temperature, volume and pressure of gases vary. From their results they produced laws about how the gases behave when you hold one of these three things constant, vary one of the others and observe the third. Together they are known as the combined gas laws. They fitted with, and acted as a proof of, Bernoulli's kinetic theory of gases.

Everyday examples of these laws at work include a bicycle pump getting warm as you compress the air in it or the interior of a fridge getting cold as the gas in the pipes wiggling around its back is made to expand. Also, burning petrol vapour in car engines' cylinders causes its temperature to increase which make its pressure and then the volume increase which pushes the pistons and makes the car go.

All we need concern ourselves with here is that as you compress a gas its molecules are made to fly about faster, and this is in effect a rise in temperature. The piston in a pump which is compressing a gas is like a table tennis bat hitting a ball. When it hits the ball the speed of the bat is added to the speed of the ball and it returns faster. In the same way the molecules return faster as the piston 'returns' them.

When a gas expands, its molecules slow down. This is like a ball hitting a table-tennis bat that is being pulled away from it. The speed of the ball is reduced. A table tennis player's drop shot makes a ball that was flying fast towards their opponent fall just the other side of the net. This is also how refrigerators keep your beer cold.

Things get colder as their molecules slow down. So just how cold can they get? Well, if molecules keep slowing down, eventually they will stop. Things can't get any slower than stopped so anything whose molecules have stopped have reached a temperature below which it is not possible to go. Absolute zero -273 degrees centigrade[2].

So that's what temperature is. It is the speed at which an object's molecules are moving or vibrating. The faster they are moving the hotter the object is. The molecules of a gas at 0°C are moving at about 1000 mph. Heat energy (what normal people call heat) is a measure of not only how fast an object's molecules are moving but also, how many it has and how heavy they are. It takes more energy to get a lot of molecules moving faster. It is easier, cheaper and quicker to boil a kettle holding just a cup of water than it is to boil one full of water. A gallon of water at 50°C has more heat energy than a teaspoon full of water at 50°C. The bigger and/or hotter a thing is the more heat energy it has.

You may be thinking that if gas molecules are flying around at 1000 mph the speed of the bicycle pump piston at much less than 1 mph is not going to have much effect. The thing is the molecules bounce off the piston millions of times a second and so the speed of the bicycle pump piston is added on millions of times a second.

So, heat energy is the kinetic energy an object's molecules have because their molecules are vibrating or flying about. Near the end of *Chapter 6 - Atom part 1* we see how Albert Einstein proved that molecules exist, and that their speed is in effect their temperature.

We are stuck with the energy we have

Returning to the cliff, the rock had a lot of potential energy when it was 10m above the car because it was in a position where the force of gravity could pull it 10m towards the car, accelerating all the way. By the time it hit the car it was moving fast and had a lot of kinetic energy. The potential energy it had to start with had been converted to kinetic energy while it was falling. Then the damage happened. The rock exerted a big force on the roof of the car and a dent formed. Is that the end of this little story? Was the energy "used up"?

---

[2] At absolute zero molecules will still be moving a tiny amount because the Uncertainty Principle tells us that there is a limit to how well you can know the momentum and location of a particle. The better you know one the less well you know the other. The Uncertainty Principle will be introduced in *Chapter 10 – Atom part 2*.

Not quite and no. In 1798, Count Rumford, a British/American scientist working for the Bavarian Army was making cannons by boring large holes in large lumps of brass. He noticed how hot the brass became when he was boring and, if he did this to a lump of brass immersed in water, he found he could make the water boil just by boring. Maybe this was the first time the people had seen water boil without using fire. He made measurements and found the kinetic energy put into the boring machine was equal to the heat energy the brass gained as it warmed. So, what Count Rumford had shown was that the kinetic energy put into boring the cannon had been converted to the heat energy the brass gained.

Throughout the 19th century James Joule, Julius von Mayer, William Grove, Sadi Carnot, William Rankine and others did further experiments like this, turning energy from one form to another. They all found that the amount of energy they put into their experiments matched the energy at the end of the experiment. Thus, the law of conservation of energy was established.

This means energy cannot be created or destroyed. It can only be converted from one form to another. The universe currently has the same energy it had when it got started. And this is the amount of energy it will always have.

This, however, does not fit well with the way we talk about energy in normal life. We say things like "I've used up all my energy", "this power plant produces a lot of energy". From a science point of view, you might say "I've converted all of the potential energy in my food to kinetic energy", or, "this power plant converts a lot of chemical energy into electric energy".

You may have heard talk of mass being converted to nuclear energy which seems to belie this. Fear not, this will be dealt with in *Chapter 12 - Nucleus*.

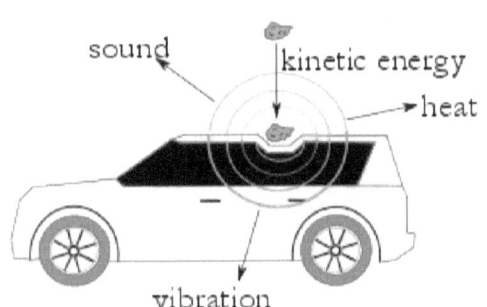

So, if energy is always conserved, what happened to the energy after the rock hit the car?

It got converted into sound, vibration and heat.

Sound and vibration are also types of kinetic energy. They are vibration of molecules in air and solids. So, in this way sound and vibration are not so different from heat.

It is not hard to imagine that the rock hitting the car would cause sound and vibration but what about the heat?

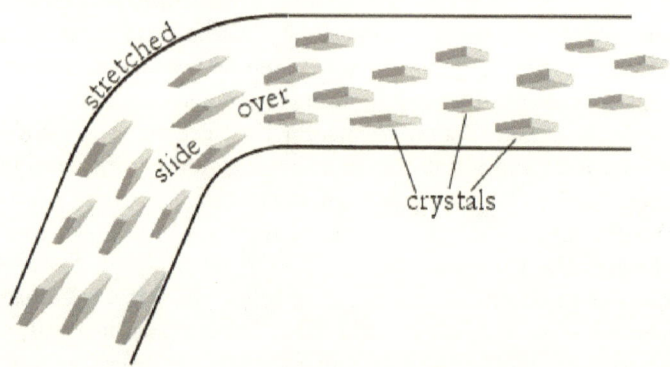

Fig 1-12 Edge view of bent metal with exaggerated crystals

The rock caused the metal in the car's roof to bend. When metal bends the surface on the outside of the bend stretches. This causes the tiny crystals that make up the metal to slide over each other. These crystals have jagged shapes which means they crash about making the molecules in the crystals vibrate faster. So, they get hot because of the friction first documented by Abu Al Biruni.

Energy is also conserved for the person on the swing. When she got on the swing, she used energy from the food she had eaten to create kinetic energy as she leaned back and forth to make the swing go. When she stopped moving her swings would reduce in size until eventually she stopped in the middle. Her kinetic and potential energy had been converted into heat energy by the friction of going through the air and the parts of the bearings sliding over each other at the top of the swing, both of which will have warmed up a bit. The energy in the person's food eventually went into slightly warming the swing and surrounding air.

While the dent was forming in the roof of the car kinetic energy was being converted into sound, heat and vibration energy. The dent got deeper until all the kinetic energy had been converted. In fact, you could use the depth of a dent to measure kinetic energy. Additionally, whenever energy is converted from one form to another, a force acts. And conversely, every time a force acts, some energy is converted from one form to another.

Thermodynamics

The realisation that energy is always conserved, and never created or destroyed, became the 1st law of thermodynamics.

The 2nd law of thermodynamics is the acceptance that heat flows from hot things to cold things and never in the opposite direction. If you put a hot potato and an ice-cream in an insulated box, the ice-cream will melt, and the potato will cool. Heat flowed from the potato to the ice-cream, so it would be sensible to keep them separate. This seemingly obvious statement has some consequences.

It implies the universe will end in a 'heat death'. The heat in the universe will spread out until stars can no longer shine and galaxies will no longer exist. The universe will comprise cold particles spread out in space. This conclusion was reached because of another observation about what happens when things cool - the stuff involved becomes more disordered. The potato and the ice-cream became an unappetising mess. So, the law is also telling us that when heat moves between objects all the objects involved end up in a more disordered state. But don't worry about heat death. It is so far in the future that we don't have a name for a number that big.

It would be natural to object and say that when a house is built this comes from clay and other substances that were dug up from the ground. The house is more ordered than the stuff that was dug up from the ground. However, the process of building a house requires a large amount of energy to extract and process the materials, transport them to site and lay the bricks. The amount of order in the house is less than the amount of disorder created by the building process in the environment as a whole.

Scientists have a name for the measure of disorder, they call it entropy and a more formal statement of the 2nd law says that entropy always increases. The mathematical formulation of this helps engineers to design better engines for our cars and planes.

A more acceptable manifestation of the 2nd law is that you will never see broken bits of a mug of tea spontaneously gather up its contents from the floor, join together to form a mug and jump back up onto a table. In this way the 2nd law is telling us that time can only go forwards. Sorry 'Back to the Future' fans.

The effect of power

We have seen that when a force acts some energy is converted from one form to another. This is usually from some form of potential energy to some form of kinetic energy. The strength of the force created depends on how fast the energy gets converted between the two forms.

If you are designing a car and you want it to accelerate faster than a competitor's product, you will want the engine to deliver a force which is stronger than your competitor's. You will want the engine to convert energy from the potential energy in the fuel to the kinetic energy of the car at a faster rate. This rate of conversion of energy is important to engineers so they have given this a name: power.

So that's why if you take a car and put a higher power engine in it, it will accelerate faster but will also use the fuel up faster as it converts potential energy to kinetic energy at a faster rate.

Types of energy

So, we have seen 4 types of kinetic energy: a moving object (regular kinetic energy), sound, vibration and heat. In fact, the last 3 are particles moving within some object or substance.

There are also different types of potential energy. One is an object in a high place. This could be called gravitational potential energy, but people just simply say potential energy for this. A battery provides electrical potential energy in the form of charged particles that are held separated from each other. This too has similarities with a rock in a high place.

Another form of potential energy is a tank of hydrogen or a plate of food. This is a form of chemical potential energy. This all seems very different from a rock in a high place but when we look at it in detail, we find it is very similar. And in fact, chemical potential energy turns out to be electrical potential energy in disguise. All will be revealed in *Chapter 11 - Chemistry* after we have looked at atoms and electricity.

Finally, there are two types of nuclear potential energy (weak and strong) and even these are like a rock in a high place as will be seen in *Chapter 12 - Nucleus*.

Each of these four types of potential energy (gravitational, electrical, weak nuclear, strong nuclear) has its own type of force. In later chapters we will see how those forces and the potential energy associated with them flowed from the Big Bang (*Chapter 14 – Cosmology part 2*) into the molecules and atoms around us and in us.

As a final look at what energy is, here is a list of some of the machines we use to make life more pleasant as they convert energy from one form to another:

- Machines that turn potential energy into kinetic:
- Petrol engine in a car
- Electric motor in a washing machine
- Central heating
- Gun (does not make life more pleasant)
- Guitar

Machines that turn kinetic energy into potential:
- Electric generator
- Microphone

Machines that turn potential energy into potential energy of a different type:
- Hydroelectric dam
- Gas power plant
- Nuclear power plant (*Chapter 12 -Nucleus*)
- Radioisotope generators (*Chapter 12 - Nucleus*)

So, energy is stuff moving (kinetic) or stuff in a place where a force could make it move (potential). It is the only thing that has the ability to supply a force. When hearts beat, when sounds are made, when cars move, they do so because a force has acted. And the force has acted because some energy has changed from one form to another. And although energy can change from one form to another, the amount of energy in our universe is fixed for all time.

Understanding energy has brought comfort and enjoyment to our lives and provided a bedrock for a deeper understanding of all fields of physics as will become apparent in later chapters. It seems like pretty nebulous stuff. You can't see it or smell it but at the start of the 20th century a brilliant patent clerk was to show that it is also responsible for the existence of mass.

Key points of this chapter
- Gravity is a force of attraction between any two objects with mass
- Newton's laws of motion and gravity apply on earth and in space
- There are two main forms of energy: kinetic and potential
- A moving object has kinetic energy
- An object that is subject to two opposing forces has potential energy
- When a force acts an object changes speed and or direction
- Forces act only when energy transforms from one form to another
- The amount of energy in the universe is constant
- Temperature is a measure of the speed of molecules
- Heat is the kinetic energy of the molecules of an object

2 CHARGE IN SOLIDS

You have probably experienced the situation where you unpack a new appliance and the polystyrene beads cling to the polythene bag. This effect is due to static electricity, which, along with its sister magnetism, was first documented over 2000 years ago.

In this chapter we will see how electricity and magnetism were discovered, how we gained enough understanding to make use of them and how they laid down the bedrock for solutions to mysteries in atomic physics, chemistry and ultimately, life. In later chapters, we will see how the electric force provides the punch behind the atomic bomb.

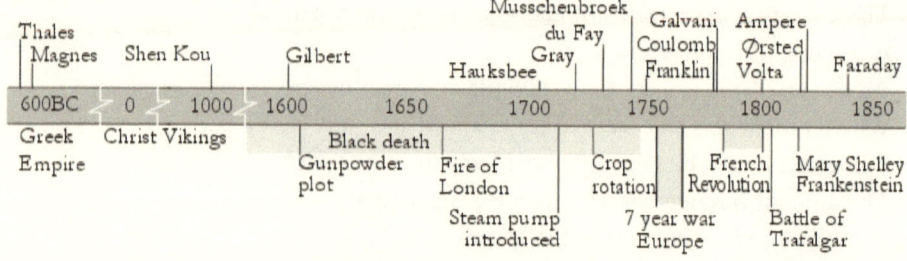

First clues of electricity and magnetism

Around 600BC a Greek philosopher named Thales accidentally started the journey to our understanding and control of electricity when he noticed little bits of straw jumping up and sticking to a piece of amber he was polishing. To us this effect is just a mild irritation when we apply cling film and it sticks to itself, but for people in ancient Greece it would have seemed like magic.

At about the same time legend has it that a shepherd named Magnes was walking through part of Magnesia in Greece and felt the nails in his sandals being attracted to a rock, an effect we now call magnetism. These pieces of rock that attract pieces of iron were given the name lodestone.

The reason for these effects remained a mystery and it deepened in the 1720s when Stephen Gray, a pensioner at Charterhouse in London, experimented with static electricity using a glass tube which had a cork in one end. When he rubbed the tube with silk, he found that small pieces of paper were attracted to the glass *and to the cork*. The effect had spread from the glass to the cork. To satisfy his curiosity he attached a stick to the cork and tested to see if the pieces of paper would be attracted to the stick. They were. It still worked when he tried an 800-foot length of metal wire. However, if the wire touched the ground the pieces of paper fell off. Conversely, if the wire was suspended by silk the pieces of paper stayed on longer. So, some substance that carries a force can flow through solid metals. When you think about it, it seems fairly magical. The answer to how it happens was not found until the start of the 20th century as we will see in *Chapter 10 - Atom part 2*.

Gray had discovered electrical conductors and insulators. Electrical conductors are objects made from a material that electricity can move through, such as metal wire or the body of a car. Insulators are objects made from material which electricity cannot move through, such as silk or plastic. In *Chapter 13 - Atom part 3* we will see why some materials are good at conducting electricity while others are bad. Later in this section we will see what caused the pieces of paper to fall from the conductor when it touched the ground.

Today electrical conductors are strung between pylons across the land to bring electricity from power plants to your house, through the walls into your computers and into its chips which do their stuff so you can play Candy Crush. And while they're at it, conductors go into your toaster and light bulbs, so you can enjoy your toast and find it at night. These conductors are mostly copper because it is fairly cheap and fairly good at conducting electricity. Silver would be a bit better, but it is not usually worth the cost.

Gray's experiments showed that:
- The effects of static electricity occur after an insulator has been rubbed with something soft
- When one part of a conductor gains static electricity, it spreads out throughout the whole conductor
- Some materials are better at conducting electricity than others

These phenomena were only explained after many layers of science had been understood. It was amazing that this electricity stuff could flow through solid material, an effect that was only explained when it was realised that electricity is part of all matter. This will become clear as we look at discoveries about the nature of matter in later chapters.

Two poles of magnetism

In Qin dynasty China (221-206BC) people had found that if a lodestone were suspended by a thread it would mysteriously align itself in a north-south direction. The first magnetic compass was invented, a great boon to navigators and the start of the scientific trail that uncovered the riddle of magnetism and other seemingly unrelated aspects of science.

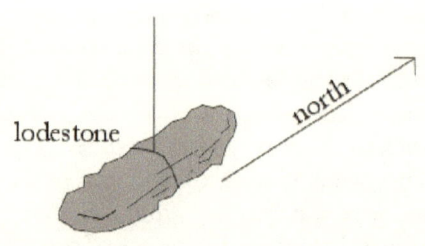

Later it was found that iron needles could be made to take on the properties of magnetism by stroking them against a lodestone to make better compasses. Also, it was found that larger pieces of iron could be magnetised in this way to produce what we would call magnets.

Fig 2-1 Magnetising a needle by stroking it against a lodestone and using it as a compass

Further insight into magnetism came in the 16th century when the doctor to Queen Elizabeth I of England, William Gilbert, played about with a pair of magnets. He found that if he put them near each other one way round they snapped together but if he reversed one of them, they repelled each other. This led him to the conclusion that there is something special about the ends of the magnets and this would explain why magnetic compass needles point north. He named the ends of a magnet poles because the ends of magnetic compass needles are attracted to the poles of the earth. And he came to the conclusion that the earth is itself a giant magnet whose north pole attracts the south pole of compass needles and whose south pole attracts the north pole of compass needles.

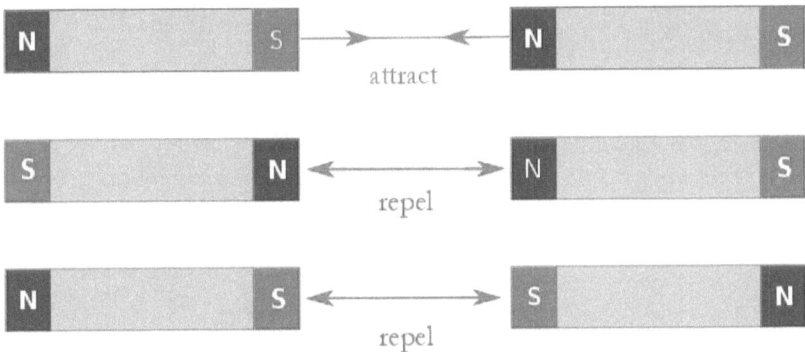

Fig 2-2 Like poles repel, unlike pole attract

What he found can be summarised as:
- Like poles repel
- Unlike poles attract
- The force between magnets is stronger when they are closer together
- The earth is itself a gigantic magnet

This raises the questions 'why is the earth a giant magnet?' and 'why do the magnetic poles of the earth coincide with the point about which the planet rotates?'. Later we will see this is not a lucky coincidence but a consequence of how magnetism arises.

Two types of electricity

In 1733, Parisian chemist Charles François du Fay decided to find out what materials would exhibit the effect of static electricity by rubbing various objects with fur and holding them near some small pieces of paper. He found some objects, like bricks or iron bars, produce no effect, while, of the ones that do, glass and amber produce the most pronounced effect.

He also made the following observations about objects he had rubbed:
- Two pieces of rubbed glass will repel each other
- Two pieces of rubbed amber will repel each other
- Pieces of rubbed amber and glass will attract each other
- These forces are stronger when rubbed objects are closer together

The behaviour of the two types of charge have some similarities to the north and south poles of a magnet, hinting at a link between them. That link will be described later in this chapter.

To explain these electrostatic effects, he suggested there are two types of fluid that permeate all material, and these are normally in balance and cancel each other out. But when you rub some types of material, these materials lose some of one of the fluids. Some materials lose one type of fluid and others lose the other type of fluid. He gave these fluids the names resinous electricity and vitreous electricity because one was lost when amber was rubbed and the other when glass was rubbed.

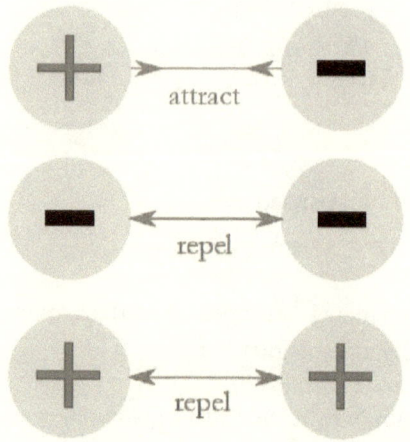

Fig 2-3 Like charges attract, unlike charges repel

In the 1740s Benjamin Franklin (one of the USA's founding fathers) performed similar experiments and came to the same conclusions. He introduced the word charge (from the French verb, charge - to load) to describe the mysterious substance that caused these effects. Also, he noticed that if you bring together two objects which have the two types of charge, they cancel each other out and you end up with two uncharged objects. This is what happens with positive and negative numbers, so he renamed the two types positive charge and negative charge. The forces between charged objects would now be summarised as "like charges repel and unlike charges attract".

Only these two types of charge have ever been found. Why are there two types? This is a question which still awaits an answer.

The electrostatic and magnetic effects can be compared to gravity (*Chapter 1 - Energy*). Like the electric and magnetic force, the gravitational force is stronger when the objects are closer together but unlike electric and magnetic forces, only one type of gravity has ever been found and it always attracts. But as yet no theory has been developed which explains the behaviour of both electricity and magnetism.

The picture formed that all objects contain a certain amount of positive charge and a certain amount of negative charge. If an object holds equal amounts of positive and negative charge it will appear to have no overall charge, otherwise known as neutral. Also, if there is an excess of positive charge the overall object will be considered as positively charged. Likewise, if there is an excess of negative charge the overall object will be considered as negatively charged.

Why uncharged paper sticks to charged glass

The above gives no indication why Thales' ordinary pieces of straw should stick to the amber he had rubbed or why Gray's pieces of paper should stick to the piece of glass he had rubbed. The pieces of straw and glass had no overall charge.

Around this time other areas of science were warming to the idea that matter comprises atoms that are too small to see. People studying the effects of charge realised the idea of atoms could be helpful to them and developed the idea of charge carriers, tiny particles that can carry charge between atoms. This explains how charge can move through a conductor and the differences between charged and uncharged objects.

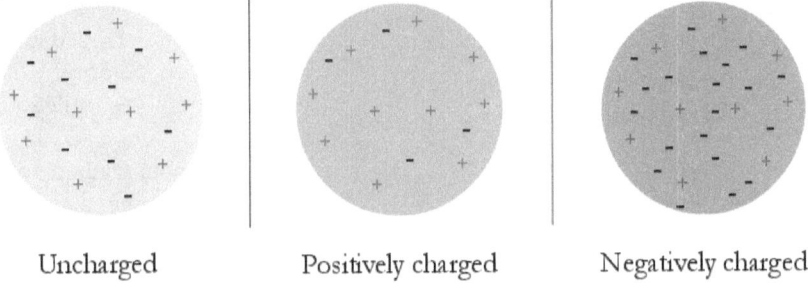

Uncharged　　　　　Positively charged　　　　Negatively charged

Fig 2-4 Charged and Uncharged objects

An uncharged object would have equal numbers of both types of charge carriers. A positively charged object would have more positive charge carriers than negative ones and a negatively charged object would have an excess of negative charge carriers.

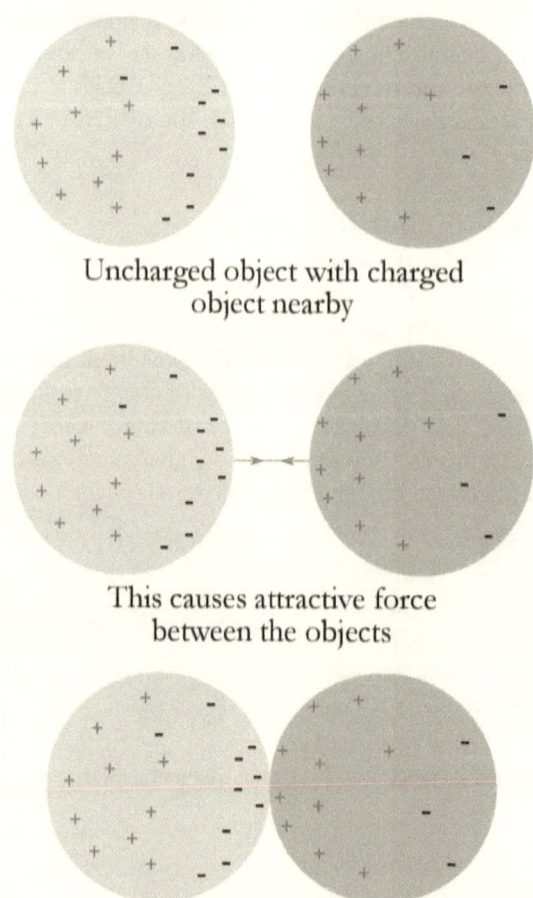

Uncharged object with charged object nearby

This causes attractive force between the objects

The two objects stick together

In an uncharged object the charge carriers will be evenly distributed across the object but if a positively charged object is brought near, the effect is like introducing a pile of Marmite™ to a room full of people. The lovers of Marmite™ flock towards it and the rest of them run to the other side of the room. You end up with a room of Marmite™ lovers on one side and haters on the other. An effect known as polarisation. A similar thing happens with positive and negative charge carriers when a positive charged object is brought nearby to an uncharged object. In the uncharged object the negative charge carriers are pulled towards the positively charged object and the positive charge carriers are repelled to the other side.

If an object with positive charge is brought close to a neutral object, the positive charge carriers experience a force moving them to the opposite side of the object while the negative charge carriers are attracted to the side nearest the charged object. And this is why the pieces of straw jumped up and stuck to the piece of amber that Thales had rubbed in 600BC and the pieces of polystyrene stick to the polythene bag when you unpack a new television.

Why lightning strikes the top of church spires

Benjamin Franklin made a second contribution to our understanding of electricity when he flew a kite in a thunderstorm and saw a spark jump to the ground from a key he tied to the bottom of the kite string. He was lucky to survive. Many people have been killed doing this to investigate lightning. Fortunately for Franklin, the kite string was not struck by lightning, it just picked up a tiny fraction of the electric charge that permeates the air in a thunderstorm. So, sparks and lightning flashes are both caused by the same

phenomenon. But Franklin also noted that the spark jumped from a sharp point on the key. He concluded that both sparks and lightning favour sharp points when choosing where to jump to and from (such as the point on the top of a church spire). This led him to invent the lightning rod - still seen at the top of most large buildings today.
Why do sparks jump from or to points on objects in preference to smooth areas? This is due to the way charge carriers spread out in a conductor.

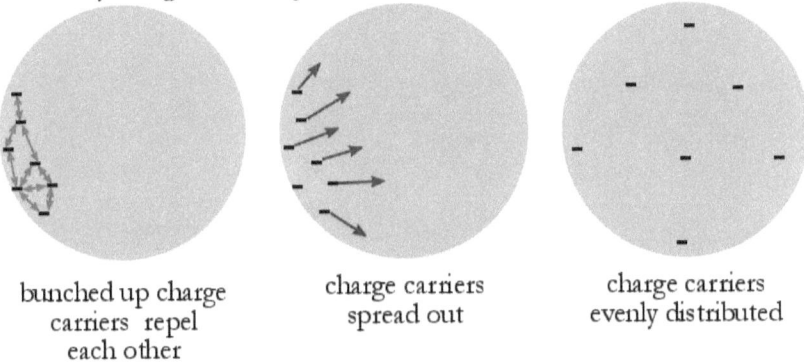

Fig 2-5 How charge carriers spread out in a conductor

If a conductor has an excess of charge carriers of the same type all bunched up in one corner of a conductor, they will all repel each other. This causes them to spread out until they have got as far away from each other as they can, after which they spread out throughout the conductor.

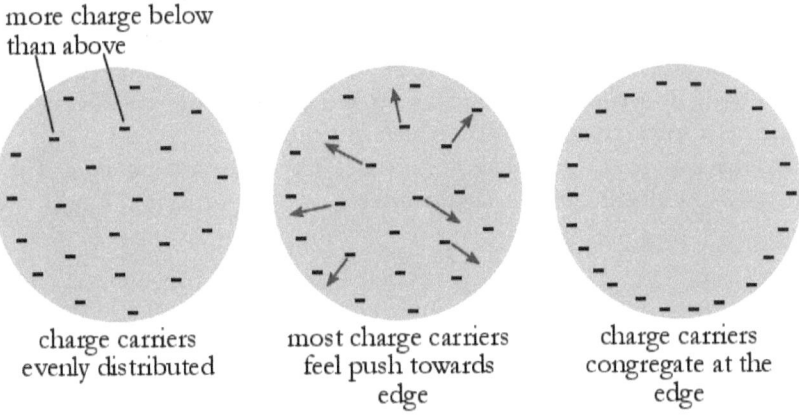

Fig 2-6 How charge carriers migrate to the edge in a conductor

Now that they are spread out evenly, consider a charge carrier a little way in from the top. There are more charge carriers between it and the bottom edge than there are between it and the top edge. This means it feels a net push towards the bottom edge. And this is true for all the other charge carriers that are not near the edge. This means they all migrate to the edge of the conductor.

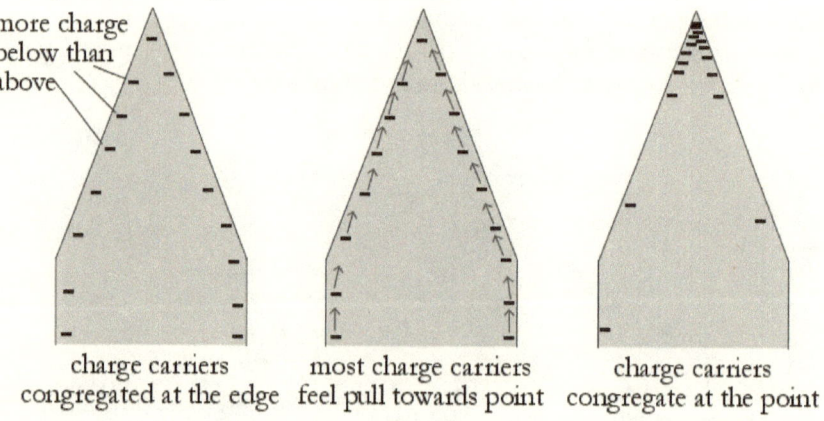

Fig 2-7 How charge carriers migrate to the point of a long spike

This effect which causes charge carriers to migrate to the surface of a conductor also causes them to migrate to the tip of a spiked conductor. The diagram above shows the pointed end of three long conductors. The one on the left shows a starting condition where charge carriers have migrated to the surface as described above. If we consider charge carriers a little way from the tip, they have more charge carriers below them than above them so in the middle diagram we see that they are repelled by charge carriers further down in the conductor and so feel a force pushing them towards the tip. And in the right-hand diagram we see they have migrated to the tip. So, the way charge distributes itself in a conductor depends on the shape of the conductor and if the conductor has a spike, charge will accumulate at its tip.

This is the first reason that lightning strikes the tip of a conductor. The second concerns how the distribution of charge affects the force it has on a nearby charged object.

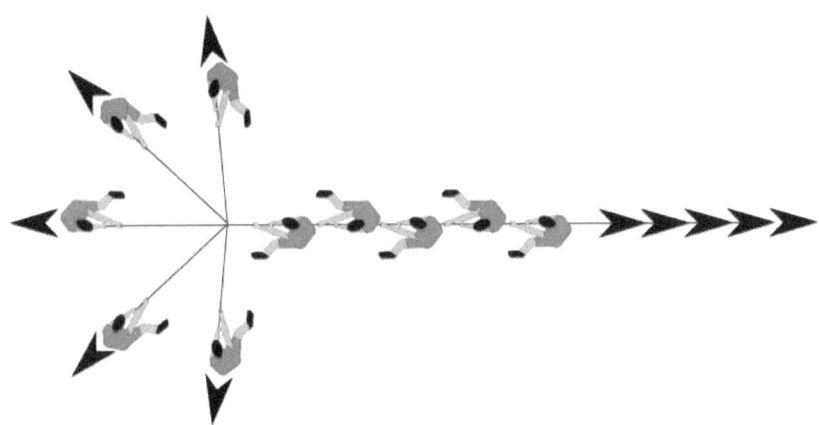

Fig 2-8 Tug of war

Imagine the charge carriers are contestants in a tug of war. One team washes regularly and uses deodorant. Team Fragrance are quite happy to form a tight group when contesting. Team Odour, on the other hand, are a bit whiffy and don't like to get close to each other. They each have a rope so they can keep apart.

Team Fragrance always wins because they are all pulling in the same direction. Team Odour has some members pulling in almost opposite directions cancelling each other out. They don't stand a chance. (They tried to get deodorant declared a banned substance, but it wouldn't wash.)

So, if a charged object gets near two other charged objects, one with charge concentrated in one place and one with charge thinly spread out, the first charged object will feel a stronger pull towards the object with the concentrated charge.

Now we can look at what happens when a thundercloud drifts over a church and why lightning is most likely to strike the tip of a church spire.

In a thundercloud, for meteorological reasons, positive charge carriers accumulate at the top and negative charge carriers accumulate at the bottom of the cloud. The effect this has on the ground underneath is to repel negative charge carriers deep into the interior of the planet and to attract positive charge carriers to the surface. Then, when the thundercloud gets near something which points upwards, like a church spire, many of those positive charge carriers accumulate at the point of the top of the spire. Also, negative charge carriers will migrate to the part of the cloud nearest to the church spire.

Fig 2-9 Charge distribution in and around a thundercloud

Now the concentrated charge at the point of the spire will exert a stronger force on the negative charge carriers in the bottom of the thundercloud than those distributed in the ground around it. So, bang, the thunderbolt goes straight to the point at the top of the spire. When this happens positive and negative charge carriers join together, and they appear to disappear. What actually happens will be described in *Chapter 10 - Atom part 2*. And because lightning is pulled more strongly to the tops of tall things lightning *does* strike twice in same place.

This is why lightning is most likely to strike the tip of a church spire.

It also explains how Benjamin Franklin's lightning rod protects tall buildings. The lightning rod is a conductor which has a sharp point at one end fixed at the highest part of the building and has the other end planted in the ground. Because copper is a better conductor than the materials used to make church spires more charge accumulates at the tip of the lightning rod than at the tip of the church spire. So, charge carriers at the base of the thundercloud feel a stronger pull towards the lightning rod than to the church spire and so that's where the lightning bolt goes. This is also why a spark is most likely to jump to or from a point on an object.

The same thing is happening but on a smaller scale.

What we have seen here also explains why the bits of paper fell from Stephen Gray's wire when it touched the ground. The wire had an excess of negative charge which caused the pieces of paper to polarise then stick to his wire. And when the wire touched the ground the excess positive charge in his wire could spread out into the planet earth which is very big compared to the wire, so the wire no longer had an excess of charge and so the pieces of paper fell off.

When engineers are worried about unwanted sparks, they make their conductors ball-shaped because a ball has no corners or points. If you are caught in a thunderstorm the advice is to put your feet close together and crouch down to make yourself like a ball.

In this section we have seen:
- That charge has a stronger pull if it is denser
- That charge carriers migrate to the surface and the tips of spikes in conductors
- Why lightning is attracted to the tops of tall buildings and tips of lightning rods

These effects also cause atoms to bond together and are the root cause of chemistry as will be seen in *Chapter 11 - Chemistry*. Without this effect there would be no chemicals and we would not be here. By the term bond all that is meant is that there is an attractive force between two or more particles, so they stick together. There is no physical 'glue'.

Ways to detect and store charge

When testing to see if an object is charged you could put some small pieces of paper near it to see if they jumped up and stuck to the object. This is inconvenient. Alternatively, you could hold the object of interest near a pointed object which is connected to earth to see if a spark jumps from it. This has the disadvantage that you lose the charge you are investigating.

To avoid these problems a device called an electroscope was invented. Various designs were tried, starting with a pivoted needle made by William Gilbert. The most successful version comprised a pair of very thin pieces of metal foil fixed to each other at one end. When a charged object touched on the ball at the top, charge spread through the conductor underneath it and into the two pieces of gold foil. Because like charges repel, the two pieces of gold foil separate, and the amount of separation is an indication of the amount of charge. This is similar to what happens when someone touches a charge generator and their hair stands on end. The pieces of foil are held in a transparent box to eliminate the effects of air movement.

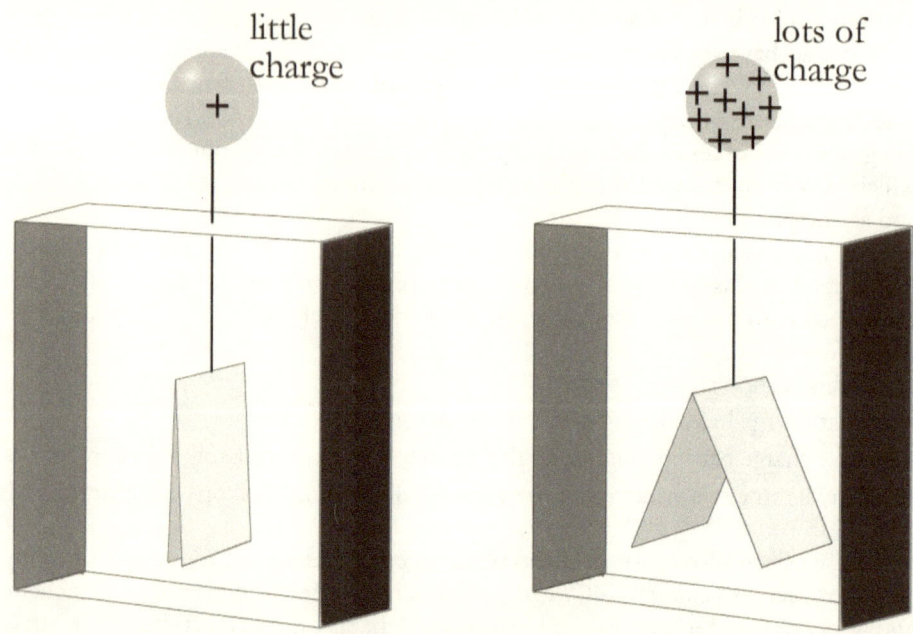

Fig 2-10 Electroscope with and without charge

Electricity investigators also wanted a more convenient way to generate charge than rubbing insulators with a cloth, and several people produced mechanical devices to do this. A popular one was produced by British scientist Francis Hauksbee who created a device which had a glass ball which was made to spin fast via pulleys and a hand crank. A piece of woollen cloth rubbed against the ball causing the ball to become positively charged.

As well as a way to generate charge, investigators of electricity wanted a way to store and carry it about. In 1746 Dutch scientist Pieter van Musschenbroek invented a charge storage device based on a glass jar. He put a layer of metal foil on the outside, another layer inside and arranged for a metal conductor to touch the inner foil. Once he had charged it by using a device like Hauksbee's charge generator he touched the terminal (a convenient place to make an electrical connection to a device) and the outer foil and realised he had stored a large amount of charge. He became the first person to experience and document an electric shock. This charge storage device became known as a Leyden jar after the town in which it was invented.

So why should this arrangement of conducting foils allow the storage of so much charge?

To answer this, we need to look at what happens after the inner foil receives the first bit of positive charge. Negative charge carriers in the outer foil are attracted towards it but they cannot reach the charge that is attracting them because they are stopped by the glass of the jar. They stay where they are, feeling the pull of the positive charge on the other side of the glass. Positive charge carriers in the outer foil are repelled by the charge on the inner foil and spread out into the table the Leyden jar is standing on and then into planet earth. As more positive charge is supplied to the inner foil more negative charge is attracted to the outer foil from planet earth via the table. This goes on and on until there is a large amount of charge on the two foils.

Now, when the Leyden jar is removed from the table the two types of charge in the two foils attract each other and keep each other in or on the jar (the positive charge supplied by the charge generator and the negative charge borrowed from planet earth). They are trapped by their mutual attraction until someone touches one end of a conductor on the terminal and the other on the outer foil.

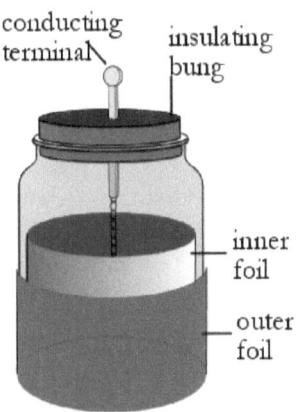

Once it was established how Leyden jars worked it became clear that all you needed was two flat conductors close together but not touching and this is essentially what modern-day capacitors are. Capacitors exist in every electronic device and are used by engineers whenever they want to store some charge for a short time.

Now there were two ways to make charge disappear: touch negatively and positively charged objects together or let either of them touch the ground. Later chapters will cover the discovery of what these charge carriers are, and this disappearance of charge will lose its mystery.

The first quantitative rule of electricity was found in 1783 when French scientist Charles Coulomb formulated a law which predicted how the force between two charged items increases as the distance between them decreases. This relationship has a lot of similarities with the way the gravitational force varies between two objects. In both cases the force is stronger when the objects are closer together. This tends to make people think these two forces may have something in common. But, even today, we have no verified theory which describes both the gravitational force and the electric force. Coulomb's name is now used as the unit which describes how much charge an object has.

We have seen in this section how devices to detect and store charge were developed and how they are based on Franklin's observation 'like charges repel, unlike charges attract'. But these devices which detect and store stationary charge have no effect on our lifestyles. In the next section we turn our attention to charge moving en-masse. This has transformed our lifestyles.

Frogs' legs led to batteries

In the second half of the 18th century, Italian scientist Luigi Galvani was watching his assistant dissect a frog's leg. The scalpel touched a brass hook that was holding the leg still. When this happened, a tiny spark was seen, and the frog's leg twitched. Galvani concluded the charge had been created by the frog's leg itself. It is not too surprising that he would come to that conclusion because at this time the capabilities of electric eels and torpedo fish were known and may have influenced Galvani. Or maybe he was wedded to the notion of a 'spark of life'. But what Galvani had shown was that nerves which control muscles and send information to the brain do so by the conduction of charge. He thought the frogs' legs had generated electricity just like electric eels do (wrong), but he was right about nerves conducting electricity.

This led him to carry out other experiments with frogs' legs and electricity. He was probably influenced by Benjamin Franklin's experiments with kites charging a Leyden jar in a thunderstorm. He connected a wire suspended in air to a frog's leg and noticed that it twitched during a thunderstorm giving more evidence that electric charge is an important part of muscle movement.

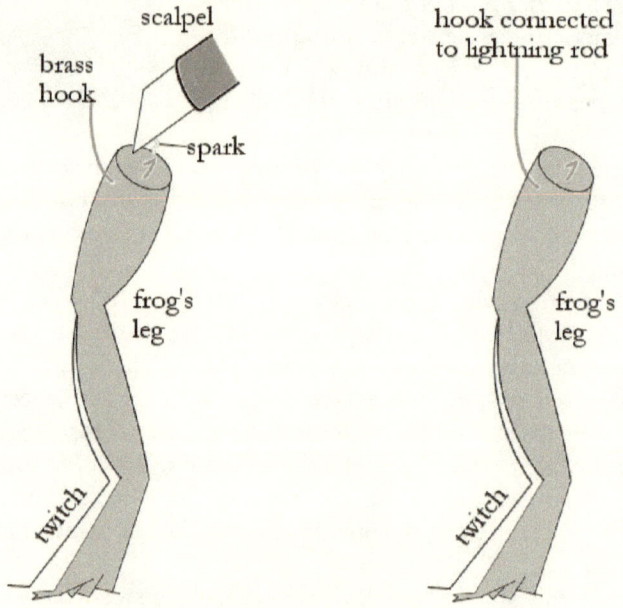

Fig 2-11 Electrical effect on frogs' legs

When news of this reached Mary Shelley, it provided inspiration for her to write Frankenstein, probably the first science fiction novel. Galvani clung, however, to his belief that the charge had been created by the frog's leg.

In 1800, Italian scientist Alessandro Volta (not the French philosopher Voltaire) wondered if the twitching of Galvani's frog's leg was to do with the fact that the steel scalpel touched the brass hook that was holding the frog still. He carried out his own experiments and found he could make a frog's leg twitch by touching certain parts of the leg with two probes that were made of different metals. He concluded that the arrangement of two different metals separated by the frog's leg created an electric charge which caused the leg to twitch.

He tested this by replacing the frog's leg with his tongue. He experienced a bitter taste when the two probes touched. This only worked if the probes were made of different metals and were wet. If they were made of the same metal, there was no effect and no bitter taste. (Don't try this yourself.!) Volta had shown that charge was generated by the two dissimilar metals and this caused the frog's leg to twitch.

Next, he replaced his tongue with paper soaked in salty water and the probes with discs of zinc and silver. He found he could produce different amounts of charge in this way by varying the combinations of metals and the liquid he soaked the paper in.

He found zinc, copper and sulphuric acid were a good combination. The zinc took on a negative charge and the copper a positive charge. In *Chapter 11 - Chemistry* we will see why this happens. Frogs in ponds around Italy rejoiced and Mary Shelley sold lots of books.

So eventually it was realised that three separate ways to create an electrical charge had been discovered: rubbing insulators; a wire in the air in a thunderstorm; touching two different metals together. The charge they produce is all the same stuff.

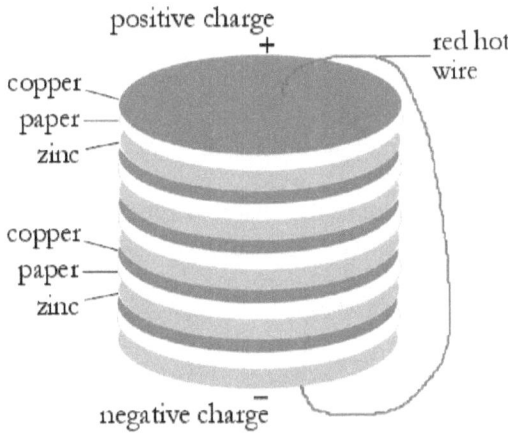

Fig 2-12 Volta's battery

After Volta had found zinc and copper was a good combination, he tried placing further pairs of zinc and copper discs separated by paper soaked in sulphuric acid on top of the first two. He found much more charge was produced. He increased the number of pairs of discs in the pile and found the amount of charge increased with each pair of discs he added.

Up until that time the best thing for holding and providing charge was a Leyden Jar which had positive charge on one terminal and negative charge on the other. If a conductor was connected between the two terminals for just a moment the charge would disappear. However, when Volta connected a conductor between the terminals of his pile of discs the charge persisted. Also, he did not need to charge his device by rubbing or waiting for a thunderstorm, the charge occurred spontaneously.

In fact, when he connected a thin copper wire between the terminals it heated up until it was red hot and then melted. This was a surprise. It was the first time something had been heated without burning fuel or friction. This was the first step to what would become the light bulb (invented by Swan and Edison late in the 19th century). Eventually the military term battery (which means a group of heavy guns) came to be used to describe Volta's pile. The wire between the ends of his battery was the world's first electrical circuit.

This was the first time that continuously moving charge had been created. Engineers are not interested in static charge because there is almost nothing of use you can do with it (apart from cling film). However, with moving charge you can heat stuff up and make stuff move (as we will see later). Continuously moving charge is the 'proper' electricity which electrical engineers use to make our lives comfortable. They use the terms Direct Current (DC) for charge that only moves in one direction and Alternating Current (AC) for charge that moves one way for a while then changes its mind and moves the other way fifty times a second. They have their reasons for wanting to make it do that, but we will not be discussing them here because AC is not important to the structure of the universe.

After Volta had invented batteries it was known that charge flowed through conductors and this fact was going to become important, but which way does it flow? They knew only one type of charge flowed because if both flowed they would cancel each other out (more of this later in the chapter). Does positive charge flow from the positive copper end of Volta's battery through a conductor to the negative zinc end or does negative charge flow from the negative zinc end to the positive copper end? The convention was established that positive charge flows from the positive terminal of a battery through a conductor to the negative one. This is known as conventional current.

But does that reflect reality? Or, to put the question another way, was it lucky that Franklin chose the positive and negative labels for the two types of charge the way round he did? Scientists had to wait until the end of the 19th century for the answer as will be seen in *Chapter 8 - Charge in Fluids*. In *Chapter 11 - Chemistry* we will see how batteries provide a constant supply of charge. The Leyden jar does not replenish charge at a terminal once it has travelled to the other terminal as the battery does. That is how a battery differs from a Leyden jar.

When Volta connected a wire between the terminals of his battery he created the first electrical circuit - a closed path that charge can move along from one terminal of a battery, through the circuit and back again. This arrangement is the basis of all electrical devices from a humble torch to the most sophisticated super-computer because it provides the constant flow of charge that electrical engineers rely on.

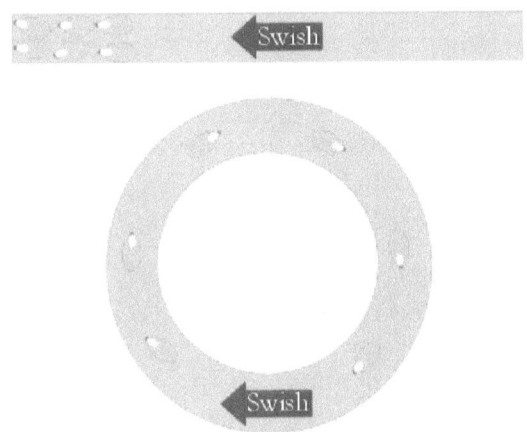

Fig 2-13 Ducks in Troughs

One way to picture why it matters that an electrical circuit forms a closed path is to imagine a trough full of water with plastic ducks floating on it and distributed along the trough. You start swishing the water in the middle of the trough in one direction with your hand. The ducks will migrate to the end of the trough but then nothing more will happen even though you keep swishing. Now imagine the trough is bent round until the ends meet and they are joined to form a continuous circle of water. Now the ducks keep moving round as long as you keep swishing. The electric circuit is like the continuous circle of water and the battery is like your swishing hand. When you switch the lights off it is like you have disconnected the two ends of the trough. This is another reason why electricity does not leak out of the socket when you remove the plug. Electricity will only flow when there is a complete circuit.

Volta's battery is like a modern-day disposable battery. It will work for a while but as it does, chemical changes happen inside and eventually it stops working. It is 'flat'. It is no longer possible to replace charge that has been removed from its terminals, unlike the battery in your phone which can be recharged over and over again[3].

---

[3] It is natural to think you could use a big capacitor instead of a battery for powering electrical

The unit Volt has been named in honour of Alessandro Volta. This familiar term is more complicated than charge. It is a measure of the force that charge in an electrical circuit feels. The more voltage you have, the more force that charge in a circuit will feel and the faster it will flow through the circuit. Voltage is sometimes called electromotive force. The force depends on the charge that is present and the way that charge is distributed. Charge is not a convenient term to use when talking about proper electricity because it is always moving about so engineers use Volt instead. This flow of charge that so interests electrical engineers is known as current.

Force fields

Until 1820 no connection between electricity and magnetism had been suspected but then Danish scientist Hans Ørsted noticed that a wire carrying an electric current would make the needle of a compass swing.

People would now say Ørsted's discovery meant there must be a magnetic field around the wire.

What do we mean by 'field'?

The word was given a scientific meaning by Michael Faraday (more of him later). For scientists it means a place where small things feel a force. In particular it is a way to describe the strength and direction of the force felt by things over some area.

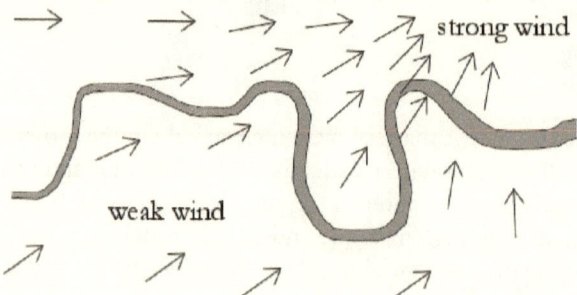

Fig 2-14 Wind field showing the forces an umbrella would experience

Weather programmes usually show a map of the country with arrows on it indicating the strength and direction of the wind in various places across the country. The direction of the arrows indicates the direction of the wind and the wind is stronger where the arrows are closer together. The weather map shows how much force you would feel on your umbrella and in what direction it would pull you. We could call this a "wind field"

---

devices. The trouble with a capacitor is that when you connect something like a light bulb to it you get a surge of charge flowing through the bulb for a fraction of a second, but it quickly dwindles to nothing. A battery provides a nice constant flow of charge that can last long enough to be of use.

as in Fig 2-14.

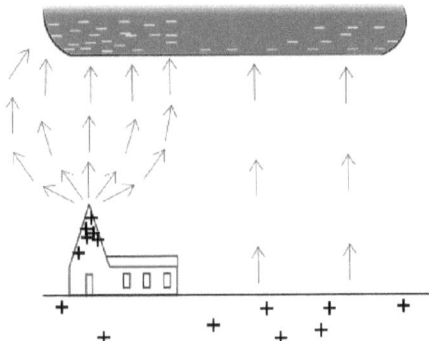

Fig 2-15 Electric field showing the forces a positive electric charge would experience

If we replace the umbrella with a positively charged object and put it in the vicinity of some large charged object like a thundercloud, we could draw a diagram of what force it would experience in various places. As described above, if a church spire was underneath the thundercloud positive charge would accumulate at its tip. If a positively charged object were placed above the church it would feel a strong pull towards the thundercloud. If it were placed next to the church the pull would be weaker. The strength and direction of these forces is represented by the arrows in the electric field diagram.

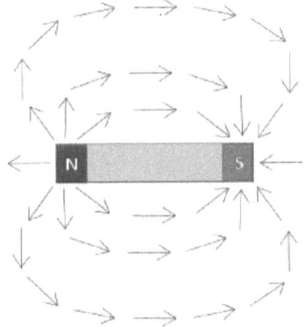

Fig 2-16 Magnetic field showing the force a magnetic north pole would experience

The arrows in the magnetic field diagram show the strength and direction the north pole of a magnet would feel if it were placed in the vicinity of one of the arrows as in Fig 2-16.

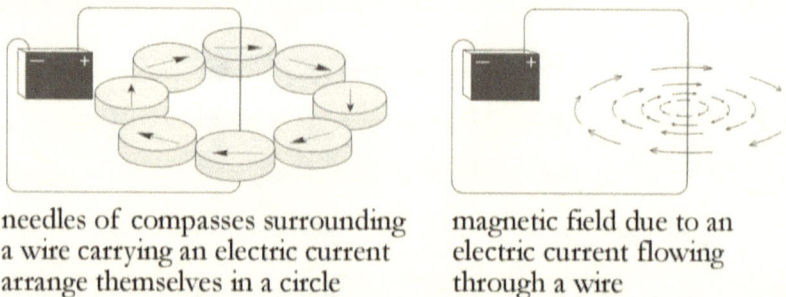

needles of compasses surrounding a wire carrying an electric current arrange themselves in a circle

magnetic field due to an electric current flowing through a wire

Fig 2-17 Magnetic compasses surrounding a current carrying wire to show the magnetic field

Returning to Ørsted's discovery that a wire carrying an electric current affects a compass - we can get an idea of the magnetic field that surrounds a current carrying wire by surrounding it with compasses. The compass needles arrange themselves in a circle indicating that a wire carrying an electric current is encircled by a magnetic field.

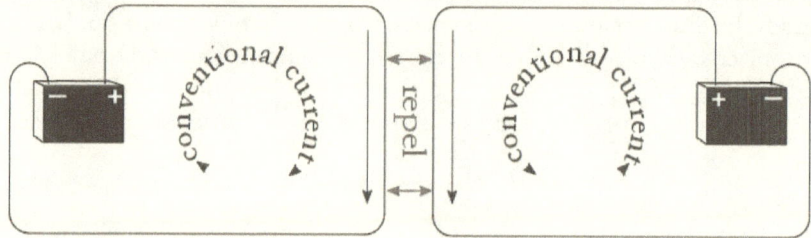

Fig 2-18 Parallel conductors with current flowing in the **same** direction

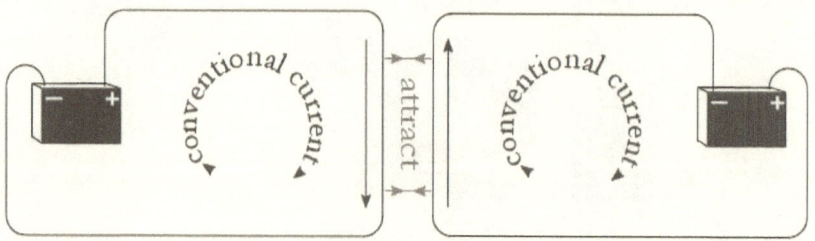

Fig 2-19 Parallel conductors with current flowing in the **opposite** direction

After Ørsted's discovery French scientist Andre-Marie Ampere showed that two parallel wires carrying a current would repel each other if the currents were in the same direction or attract each other if they flowed in opposite directions. This is because both wires create Ørsted's magnetic field and the direction of the field depends on the direction of the current through the wires. These two findings were the biggest advance in our knowledge of magnetism since Magnes felt his sandals' attraction to a rock in the 6th century BC. They show that electricity and magnetism are closely related phenomena.

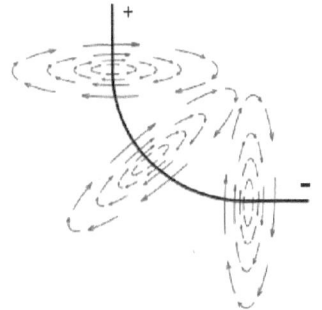

If the wire had a bend in it, it was seen that the magnetic field became more intense inside the bend. A complete loop would intensify the magnetic field further. Even more intensification could be achieved by arranging several loops of wire together (a coil).

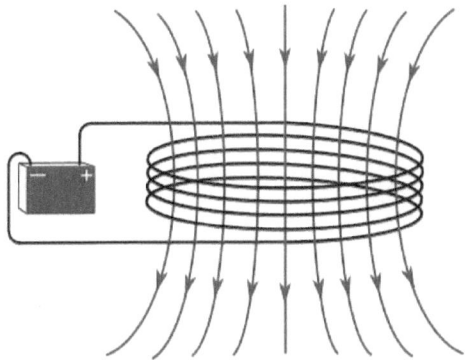

Fig 2-20 Magnetic field of a coil carrying an electric current

If a piece of iron or a magnet is placed near a coil it will be attracted to it when a current flows through the coil. This is what happens when a security door is opened. Someone presses a button which causes a current to flow through a coil. This creates a magnetic field which pulls a bolt out of the door so you can enter. This effect is also the basis for the electric motor in your washing machine. And a more efficient way to create permanent magnets.

In recognition of Ampere's and Ørsted's contributions, the unit of electrical current was named Ampere and the unit which defines magnetic field strength was named Ørsted.

In 1831 English scientist Michael Faraday thought that if a current was able to produce a magnetic field perhaps a magnetic field could produce a current. He found that it did, but only if that field was changing. If the magnetic field was constant no current was produced in the wire. This effect, known as induction, was also discovered at the same time by American scientist Joseph Henry.

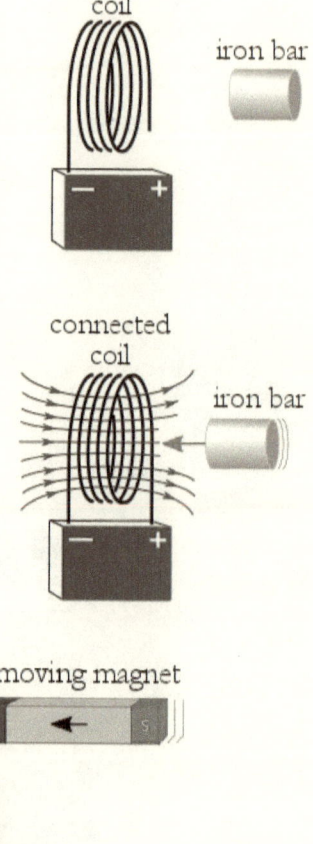

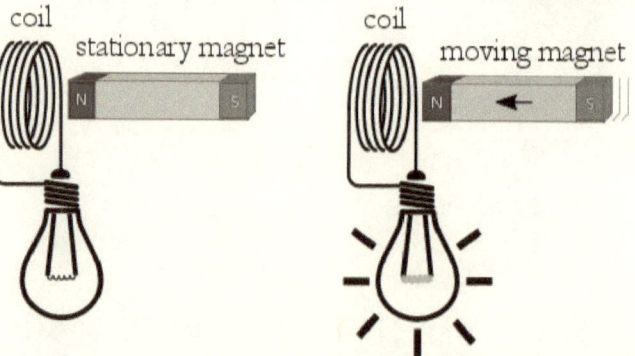

Fig 2-21 Moving magnet inducing a current in a coil

Building on these discoveries Faraday produced the first electric motor and the first electric generator. Most electric motors like the one in your washing machine work by using coils to produce magnetic fields which push the coil away from or towards a magnet. And most electric generators work by using rotary motion to move a magnet near a coil causing the magnetic field to continuously change and so a current to flow in the coil.

These scientists' discoveries along with current day knowledge of the structure of the earth provide an explanation of why the earth has its own magnetic field. The earth has a vast solid iron core at its centre which, due to its interaction with the semi-liquid layer above it, causes charge to rotate around the core and so produce a magnetic field. And because the core rotates about the same axis as the earth the poles of magnetic field coincide with the poles of the earth's rotation, a fact that has benefited navigators since Shen Kou created the first magnetic compass.

Earlier in this chapter we saw how the distribution of charge in a conductor affects nearby charge carriers. Electrical engineers use the term capacitance to indicate the physical layout and shape of conductors and the unit of capacitance was named a Farad in Faraday's honour.

We have seen that charge is something that can create a force on other charged objects and that force depends on how the charge is distributed and whether the object has the same type of charge. It can move through conductors carried by charge carriers and when it is moving it creates a magnetic field. The electric force is the one that is most responsible for making our world the way it is, from making sand stick to the bottom of wet feet to making our cars go (including petrol cars). In later chapters we will see how all this forms an important part of our understanding of the universe and life.

Why some metals stick to magnets

We saw above how an uncharged piece of paper is attracted to a piece of glass or amber that has been given charge by rubbing it. The reason some metals are attracted to magnets is rather similar to, and also depends on the idea that all substances comprise minute particles.

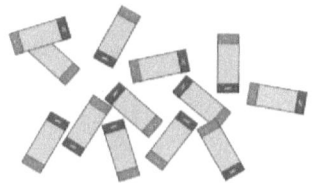

Particles in these metals act as minute magnets[4]. If we have a bunch of small magnets randomly arranged, their magnetic fields would cancel each other out and there would be no overall magnetic field. This is the situation of a regular piece of iron. If a large magnet is brought near a group of small magnets their south poles are attracted to the north pole of the large magnet. And

---

[4] The nature of these particles is outlined at the end of *Chapter 13 – Atom part 3*

the north poles of the small magnets are repelled by the north pole of the large magnet. This causes the small magnets to swing round and align themselves, so their south poles point towards the north pole of the large magnet. This in turn causes the magnetic fields of the small magnets to reinforce each other and in this way they gain an overall magnetic field. The south pole of this overall magnetic field attracts the north pole of the large magnet.

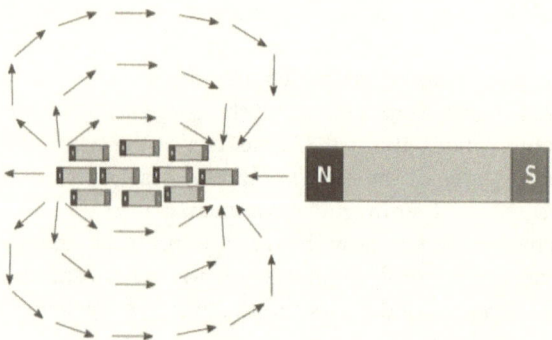

Fig 2-22 Small magnets aligned by the presence of a nearby large magnet – significant overall magnetic field

This is what happens when a regular piece of iron approaches a magnet. Minute particles within the iron align to give it a magnetic field which causes it to be attracted to the magnet. In most cases, when the magnet is removed most of the particles in the iron revert to their random orientations and the magnetic field of the iron disappears. Some of the particles may be left in an aligned orientation and by repeatedly stroking a piece of iron many more of the its particles may become aligned and retain their alignment when the magnet is removed. In this way the piece of iron retains its magnetism and becomes a permanent magnet.

Another way to align the particles in a piece of iron is to put it into a coil of wire carrying an electric charge. The magnetic field of the coil will create a permanent magnet in the same way.

The discoveries of Thales and Magnes uncovered a force of nature that has made our modern lives comfortable, fun and secure. In later chapters we will see now this force underlies most of the rest of physics.

Key points of this chapter

- There are two types of charge: positive and negative
- Like charges repel, unlike charges attract
- Because like charges repel, they congregate at sharp points of conductors which is why lightning strikes the tops of church spires
- You should crouch down if you are in danger of being hit by lightning
- Charge exerts a stronger force if it is densely packed together than if it is spread out
- Static electricity refers to charge that is not moving
- Electric current refers to charge that is moving through a conductor
- There are two types of magnetic pole: north and south
- Like poles repel, unlike poles attract
- A force field is a place where an object feels a force
- A force field is also a description of the direction and strength a small item will feel in all places in that force field
- Electric field, magnetic field, gravitational field are examples of force fields
- Electric current makes conductors heat up and gives them a magnetic field
- A changing magnetic field will generate electric forces in a nearby conductor

3 LIGHT PART 1

Light - you are probably using it right now to read this book. Since ancient Greek times people have speculated about the nature of light. The answer, when it was found, was shocking. In this chapter we will see how man's knowledge of the nature of light was built up from observing the paths it takes to an understanding of what it is made of.

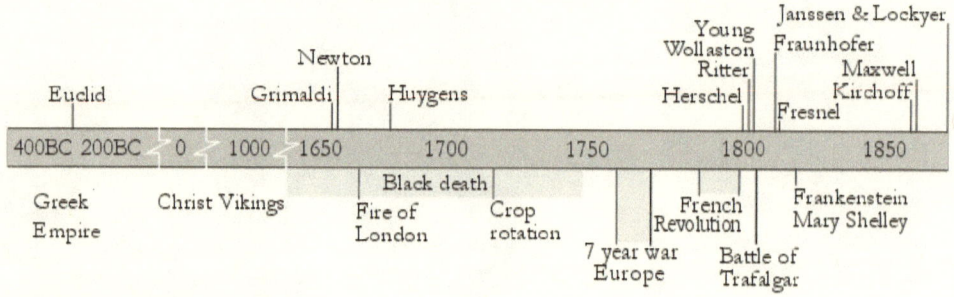

The path of light

The first insight into the nature of light came around 300 BC when Greek mathematician Euclid found that light travels in straight lines. If you line up 3 poles and look along them the first will hide the other two - a fact relied on by carpenters and surveyors between then and now.

That's about all people knew about light until 1660 when Italian priest Francesco Grimaldi decided to measure the size of a shadow. He made a beam of light fall on a thin rod and measured the size of the shadow it produced. Using the fact that light travels in straight lines he calculated how big the shadow should be. Surprisingly he found the actual shadow was wider than this and named this effect diffraction. This indicates the path of light bends when it encounters the edges of objects as can be seen in Fig 3-1.

Why should the path of light be affected by the edge of objects making shadows wider than expected?

Some people around at the time, especially French philosopher Rene Descartes and Dutch scientist Christiaan Huygens, thought diffraction could be explained if light was a waveform.

What is meant by the term waveform? It is used to describe something that is made up from waves like waves on the surface of a pond. All waves behave in a distinct way. If you put a barrier halfway across a pond and produce some waves on one side of it, they will curl round the end of the barrier. The curled part of each wave is less distinct. So, if light is a waveform, this would explain the wider than expected shadow that Grimaldi found.

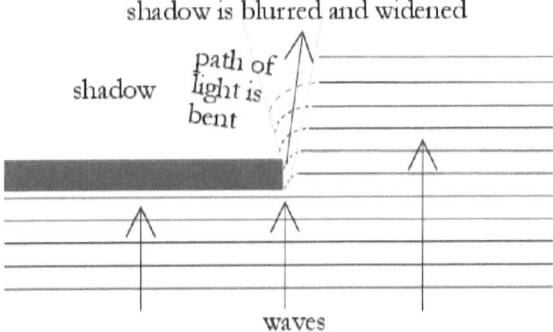

Fig 3-1 Waves passing end of a barrier

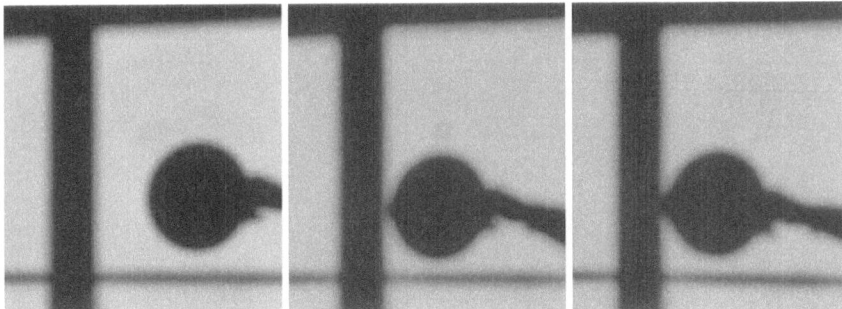

Fig 3-2 Shadow of plate showing diffraction

It is easy to demonstrate diffraction on a sunny day. Hold something round, like a plate, half way between a window and the opposite wall and look at the shadow of the window frame and the plate on the far wall. When the shadow of the plate is in the centre of the window it is round as you would expect. But as you bring the plate towards the edge of the window a bulge appears on the edge of the plate's shadow, and this bulge grows as you bring the plate closer to the edge of the window. This is due to the way light curves around the edge of the plate.

Fig 3-3 Wavelets producing the next wave

In 1678 Huygens produced a theory which explains why waves curve the way they do when they pass the end of a barrier based on something he called wavelets. The idea is that as a wave progresses every point on the wave sends out a small wave which he called a wavelet. The front edges of these wavelets combine and reinforce each other to form the next wave. Wavelets at the edge of the barrier are not reinforced on one side so the shadow spreads out.

This theory also nicely explains why waves spread out the way they do when they pass a gap in a barrier - an effect that was to become important later, as we will see.

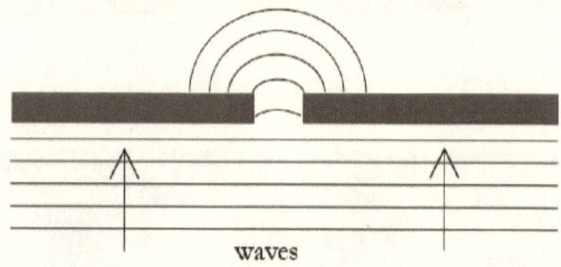

Fig 3-4 Waves passing through a gap in a barrier

So, Huygens' theory supported the idea that light is a waveform, but it also predicted that all waveforms would progress at a finite speed but gave no indication of what that speed would be. Later in this chapter we encounter theoretical work which did predict this speed.

Light spectrum

Apart from the path that light takes, people were intrigued by the colours produced when white light shone through shaped glass or drops of water forming rainbows. The appearance of colour within rainbows was well known and many people let a beam of light fall on a glass prism to investigate the colours it produced but there was little progress in understanding what was going on until 1665. This was when Isaac Newton (*Chapter 1 - Energy*) let a beam of light fall on a glass prism. But he made the crucial difference of letting the light fall onto a screen he had placed distant from the prism. This allowed the different coloured lights to spread out before they hit the screen so they could be observed.

He declared the spectrum contained the seven colours we are familiar with. The adoption of seven colours was quite arbitrary. Newton probably did this because an accepted view of the time was that seven was an important number (seven days of the week, seven planets, seven openings in the head, an idea that came from the Greek sophists). We now know there are more than seven planets and we also now consider the spectrum as a continuum of varying colour.

But how did the prism turn white light into coloured light?

Since the time of the ancient Greeks it was believed that white light was pure and coloured light was tainted in the way that red wine stains a white tablecloth. So, the consensus at this time was that a prism somehow added colour to the white light.

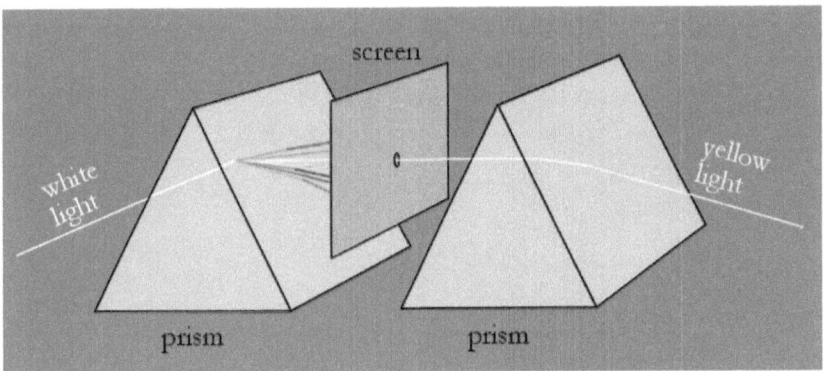

Fig 3-5 Newton's attempt to split yellow light

Newton realised he could test if this was true with the spectra he produced from prisms. "If white light was pure and the prism was adding the colours into the light," Newton thought, "Then if I select one of those colours (say yellow) by passing the spectrum through a hole then on to a second prism, then, if the theory is right, this second prism should add colours to the light and this should produce all of the colours of the rainbow again." He tried this, and it didn't. He just got the same old yellow light.

Prisms do not add colours to the light.

With a more complicated arrangement of prisms he found he could reproduce white light by combining different colours. Today a TV screen makes white light by mixing red, green and blue light.

So, Newton reached the conclusion that white light contains all colours and a prism splits white light into colours. And it does this because it bends different colours by different amounts. It bends violet light the most and red light the least.

Another conclusion he reached about the nature of light was that it is made up of tiny particles he called corpuscles. He inferred this because light travels mainly in straight lines and because when light reflects at an angle off a surface it behaves more like a pool ball bouncing off a cushion than waves bouncing off the side of a tank.

Also, this explains partial reflection. You have probably noticed your own reflection when standing indoors when you are looking out of a window on a sunny day. When this happened to Newton, he realised some sunlight had come into the room and bounced off his face, then reflected back from the glass in the window and hit his eye giving him the sensation of seeing his own reflection. But he knew that someone outside would also be able to see his face as well. So, when the light left his face some of it reflected back when it hit the glass in the window and some of it went straight through. To Newton this further supported his corpuscular theory. Some light corpuscles were going through the glass and some were reflected when they hit the surface of the glass.

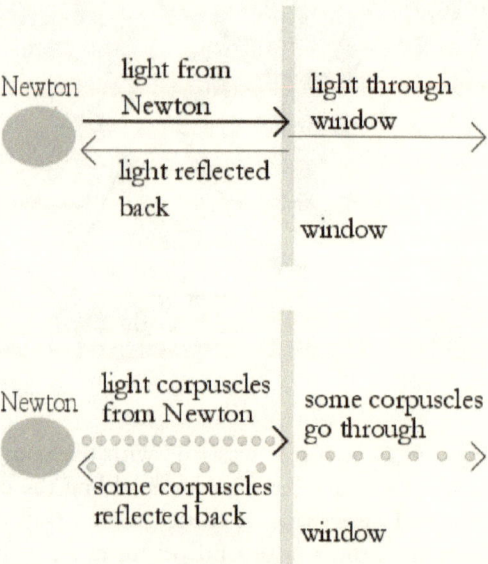

Fig 3-6 Partial reflection

So, Newton thought light was made up of corpuscles and Descartes et al thought it was a waveform. They can't all be right. Or can they? In *Chapter 9 - Light part 2* we shall see.

Light we cannot see

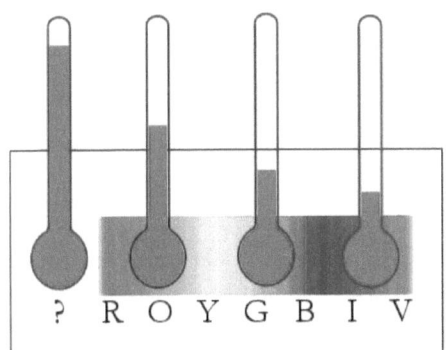

The story of the spectrum continued in 1800 when German-British scientist William Herschel let a spectrum from a prism fall on a thermometer. He let individual colours of the spectrum in turn fall on a thermometer to measure their temperatures. In this way he found blue to be the coolest and red the hottest colour. Then he moved the thermometer off the end of the spectrum beyond red expecting to see the temperature fall. He was surprised to see that it rose further. Some sort of light that he couldn't see must have been falling on it. He had discovered infrared light - an invisible light that transfers heat energy from the sun to Herschel's thermometer.

In *Chapter 1 - Energy* we saw that temperature is the measure of how fast the molecules of an object are vibrating. Also, heat spreads out in an object because if one molecule bashes into the next molecule it will make that molecule vibrate faster. If you put a cool object (piece of butter) on a hot object (frying pan) the cool one will warm up and you are ready to fry some bacon. However, objects left out in bright sunlight also warm up, but the sun is not touching them. The sun is donating heat energy to objects on earth and making their molecules move faster from a distance. In other situations, we have considered energy moving from one place to another when some object like a rock or the molecules in a warm object move and hit something. But this is different. How come the molecules in things left out in sunlight start to vibrate faster? This must mean something is carrying heat from the sun to the earth.

We use the general term 'rays' or 'radiation' for stuff that carries energy from one place to another without physical contact. Light is a form of radiation and so is the infrared that Herschel discovered. Even waves on the surface of a pond are a form of radiation. If someone moves a log up and down which is floating on a pond, waves will spread out from it. The log is being given kinetic energy which is transferred to the water nearby in the form of waves. When those waves encounter a cork floating on the other side of the pond they cause that to bob up and down. Some of the kinetic energy of the log has transferred to the cork.

Above, we saw that infrared radiation was very good at transferring heat energy from the sun to a thermometer (or anything else) on earth. Why should infrared radiation be so good at transferring heat energy? It is to do with the fact that things have a natural frequency that they most readily vibrate at. If they are exposed to vibration of the same frequency, they will start to vibrate spontaneously. If you put two guitars close to each other and pluck a string on one, the same string on the other guitar will vibrate. This is because a taut string can be made to vibrate by a tiny vibration of the right frequency and that was supplied by the string on the first guitar.

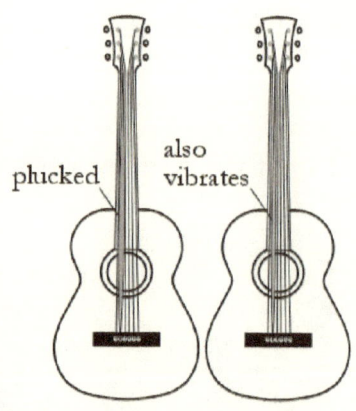

plucked / also vibrates

This is an effect known as resonance. Anything that vibrates will exhibit resonance. Another example is pendulums. This effect can also be demonstrated by several pendulums of different lengths hanging from a string. If one pendulum is made to swing, a tiny bit of the swing motion will be picked up by the other pendulums but if one of those happens to be the same length it will start to swing a large amount. Individual groups of hairs in our inner ears only resonate when they detect sound with a particular frequency. When they do so, they send a signal to the brain. Infrared radiation has a frequency similar to that at which molecules and atoms vibrate and so they readily vibrate when hit by infrared radiation. And as was seen in *Chapter 1 - Energy*, the more an object's molecules vibrate the hotter it is.

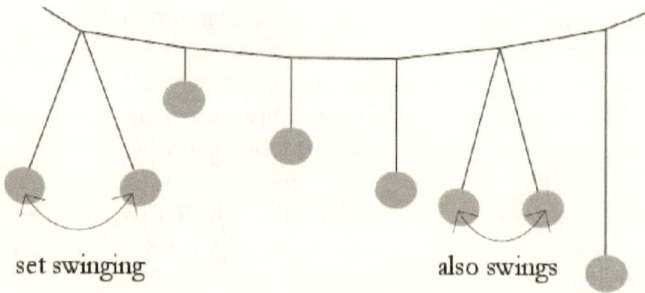

set swinging / also swings

Fig 3-7 Pendulums hanging from a string

The spectrum had been expanded to:

Infrared Red Orange Yellow Green Blue Indigo Violet

In 1801 German scientist Johann Ritter thought, "If Herschel found something beyond the red end perhaps I can find something beyond the violet end of the spectrum". He tried various things but when he placed some silver chloride paper (early photo paper) beyond the violet end of the spectrum, he found it turned black even though no visible light had fallen on it. This became known as ultraviolet light (the stuff that makes our skin tan when we lie in the sunlight).

The spectrum had been expanded to:

Infrared Red Orange Yellow Green Blue Indigo Violet Ultraviolet

If infrared light can heat an object that is some distance from the thing that is emitting it, it must be transferring some energy. In fact, when light hits our eyes some energy must be transferred to part of our eyes to cause the sensation of seeing. The basis of how this happens will be addressed in *Chapter 10 - Atom part 2*. When our skin darkens as we sunbathe it is because ultraviolet rays carry some energy from the sun to our skin. This energy causes changes in our skin which makes it darken.

Unexplained lines appear in the spectrum

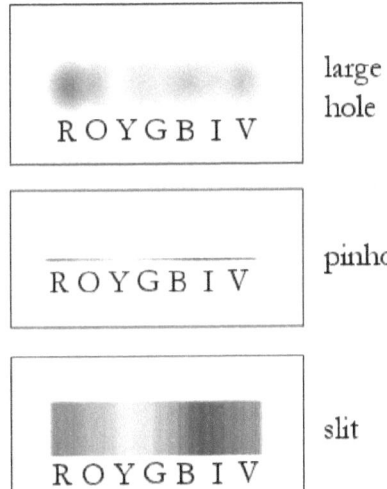

In order to produce a spectrum, people let a narrow beam of light fall on a prism. To create this beam, they would let light from the sun, or something else that was luminous, shine through a small hole in a screen. It was found that a small hole was best and producing a well-defined spectrum showing lots of colours, but it was in the form of a thin line that was difficult to see. For that reason, they started to use a slit This produced a spectrum in the form of a band that was easy to see and still showed lots of colours.

In 1802 while verifying Ritter's results, British scientist William Wollaston noticed dark lines in the spectrum of sunlight. It was realised that these dark lines corresponded to the slit shaped hole that formed the beam. Also, it meant that those particular colours were missing from the light coming from the sun.

But why should that be? This remained a mystery until the discoveries described in *Chapter 10 - Atom part 2* were made.

dark lines found in spectra by Wollaston

Wollaston could not have realised his lines were to yield so much information about the nature of matter and the universe as will be seen in *Chapter 14 - Cosmology part 2*.

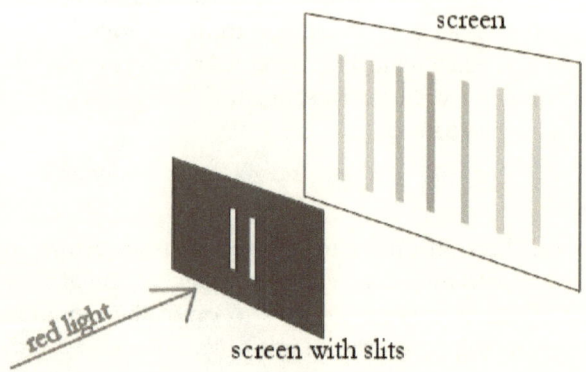

Fig 3-8 Bands of light produced by Young's slits

In the same year British scientist Thomas Young shone light through two close, narrow slits onto a screen. Instead of seeing two lines of light on the screen he saw many bands of light. This was puzzling for a while until he realised it was because two sets of waves from the two slits were overlapping to cause the pattern. Young realised this was further evidence that light is a waveform and more than that, with some geometry and maths, he could work out the wavelength of light.

What is wavelength? It is the distance between the peaks of the wave. Waves of a particular musical note will all have the same wavelength. For example, the wavelength of middle C (music) is 1.3m and the next higher C note has a wavelength half as long (0.65m). Infrared radiation has wavelengths in the range from 1 millimetre to 800 nanometres. (A nanometre is 1/1,000,000,000th of a metre). Red light is 700 nanometres and violet is 400 nanometres. Ultraviolet light ranges from 400 nanometres to 100 nanometres.

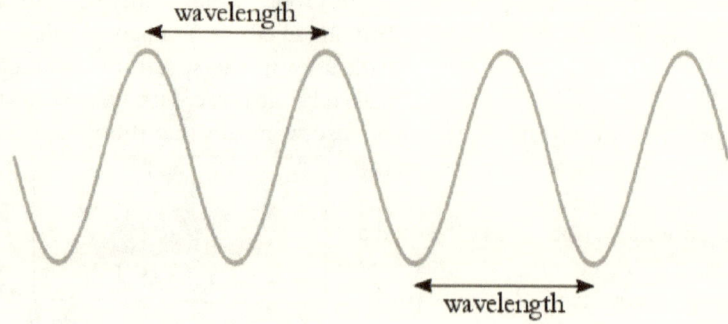

Fig 3-9 Waveform showing wavelength

In 1816, French engineer Augustin-Jean Fresnel reinforced this theory when he produced his theory of "interference" which explains perfectly why Young's two slits should produce many bands of light. The basis of this is that waves passing through a gap in a barrier will spread out as we saw above. Young's two slits were close together so when the light spread out from these two slits there is an area where they overlap.

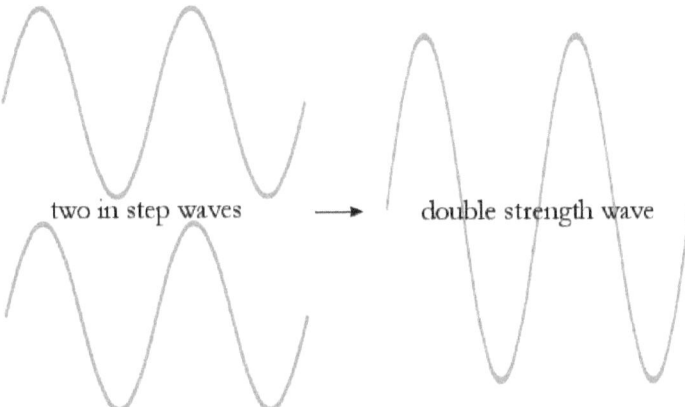

Fig 3-10 In step waves producing double strength wave

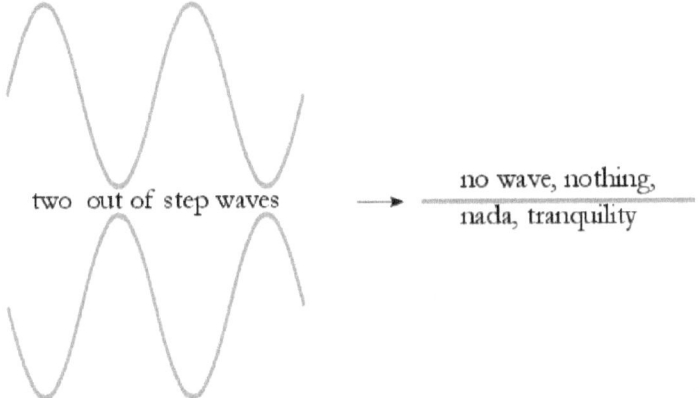

Fig 3-11 Out of step waves producing no wave

If two waves of the same wavelength encounter each other and they are in step[5], they will join forces, reinforce each other and produce a double strength wave. If they are out of step, they will cancel each other out and the result will be no wave.

Interference provides a good explanation of how the bands of light produced by Young's lists form. Referring to Fig 3-12, the light at point A on the screen had travelled the same distance from both slits so the waves are in step and they reinforce each other to make a bright band. At the points marked B one wave had travelled half a wavelength further, so the waves are out of step. They cancel each other out and that part of the screen is dark. At the point marked C one wave has travelled one wavelength further so they are in step and they reinforce each other, and a bright band appears. D, one and a half wavelengths - dark band. E, two wavelengths - bright band.

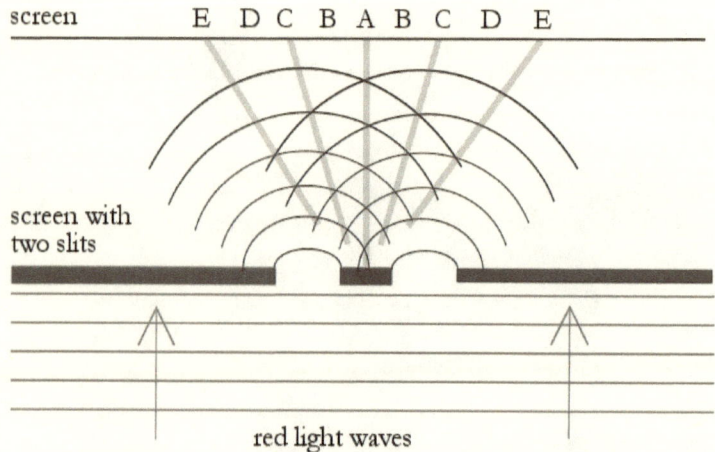

Fig 3-12 Light passing through Young's slits, viewed from above

An experiment similar to Young's slits can be done with waves on the surface of water. Send a flat wave to a barrier which has two gaps in it. On the other side, waves radiate out from each slit and overlap each other. Where this happens, you see areas where waves are high because the two waves reinforce each other and other areas where the water is calm because the two waves cancel each other out. This corresponds to the bright bands of light on Young's screen.

When Grimaldi coined the term 'diffraction' he used it to mean the way shadows spread out further than expected because of its wave nature. Now the term was extended to cover the effect of Young's slits as well.

---

[5] To be strictly correct the terms 'in phase' and 'anti-phase' should be used but the terms in step and out of step have been used here because they are more familiar.

What became clear was that each colour of light has its own wavelength and the spectrum shows how those wavelengths are distributed.

At the start of the 19th century, German scientist Joseph von Fraunhofer experimented by augmenting Young's two slits with many more slits to produce a device which he called a diffraction grating. His early versions comprised fine wires wound between rows of tiny screws with tiny gaps between the wires. This produced much sharper bands when light of one colour was used. When white light was used each band became a small spectrum.

Singling out one of these bands proved to be a better way to produce a spectrum than shining light through a prism. It yielded a much more well-defined spectrum. Later, lenses were added to this to further improve the detail that could be seen. The whole apparatus became known as a spectroscope.

The colours you see reflected in a DVD are produced in the same way that a diffraction grating produces a spectrum on a screen behind it. In fact, it is possible to make a spectroscope from a DVD with cardboard and glue to show spectra lines in different light sources. With a grating the diffraction happens as the light passes through whereas with a DVD the diffraction happens to the reflected light.

In about 1812 Fraunhofer found 600 of William Wollaston's lines in the spectrum of the sun and catalogued them. Other scientists found these lines could be reproduced on earth by shining light through gases. They also saw patterns in these lines, patterns that were unique to each gas: a barcode for gases. Because these lines are caused by gases absorbing light of certain wavelengths, they became known as absorption lines.

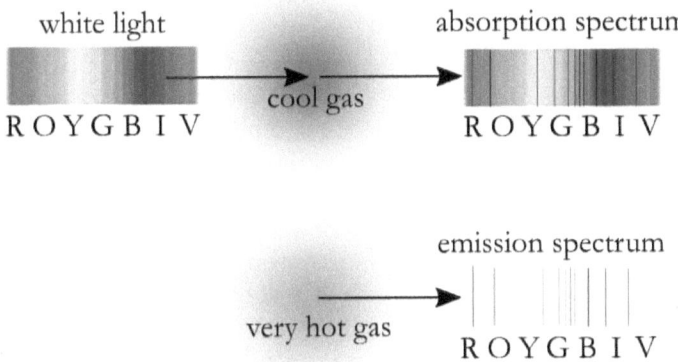

Fig 3-13 Production of absorption and emission spectra

In 1849, French physicist Léon Foucault examined the spectra of light produced by an arc which produces light from a bright spark. He found two bright lines and realised they corresponded to two of the dark lines that Fraunhofer had seen in the sun's spectrum. Following this it was found that all the patterns of dark lines found by Foucault can be produced by any very hot gas and they match their absorption lines.

These bright lines became known as emission lines. Spectra containing these lines became known as absorption spectra and emission spectra.

Emission and absorption spectra soon became a well-used technique for identifying elements in gases as will be seen later. Their cause was unknown until discoveries about the structure of the atom were uncovered as will be seen in *Chapter 10 - Atom part 2*. They were also to become the prime method for finding the speed and direction of stars and galaxies (*Chapter 4 - Cosmology part 2*).

Spectral temperature

It is difficult to compare the intensity of different colours but in the middle of the 1800s devices were developed which could measure the intensity of light. When these detectors were coupled with a spectroscope it was possible to let just a small part of a spectrum fall on the detector. In this way it became possible to measure the intensity of the light coming from an object of individual colours (and so, individual wavelengths) and then draw a graph of intensity vs wavelength.

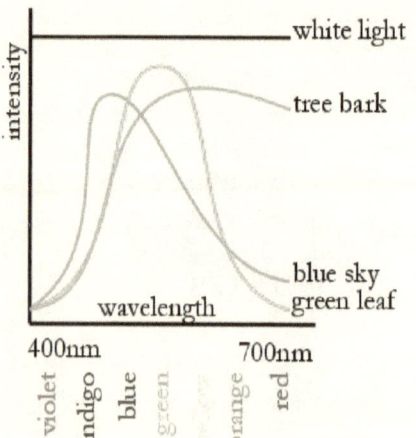

These graphs show how the colours of the things we see are a mixture of light of different wavelengths. In the graph on the left we can see that tree bark for example has very little violet light, but it does contain green, orange and red. White light has all of the visible wavelengths.

In 1862 German scientist Gustav Kirchhoff used a spectroscope in this way to produce spectrum graphs of red-hot objects.

What he found was:
- All hot objects produced a curve of similar shape (Fig 3-14)
- All hot black objects produced graphs of the same shape
- The wavelength at the peak of the curve varied with temperature

- Different objects of the same temperature produced a graph whose peak was at the same wavelength
- The curves extended into the infrared and ultraviolet parts of the spectrum

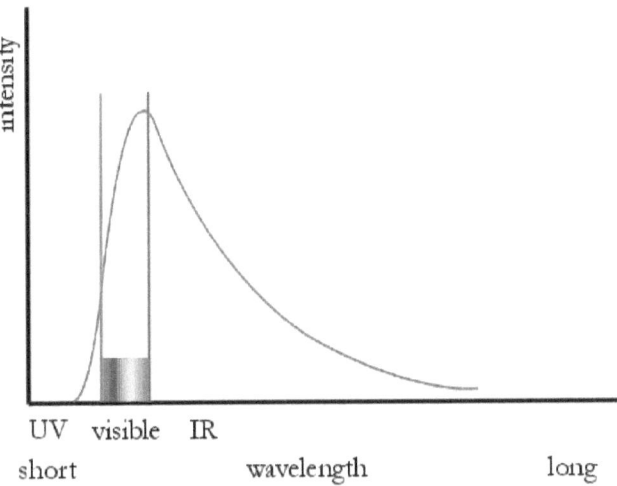

Fig 3-14 Spectral graph of an object which has been heated so it is glowing red hot

Because all objects that are black when they are cold have a curve of the same shape this curve became known as the black-body radiation curve.

As an object gets hotter the peak of its black-body curve moves to shorter wavelengths. When an object is red hot the peak of the black-body curve is at the red end of the visible spectrum. If the temperature is increased further the peak of the black-body curve moves through orange and yellow until it reaches the middle of the visible spectrum. At that point it is radiating light at all visible colours which makes it look white: white hot. Things glowing in this way include toaster elements, electric grill elements, logs on fire, coal on fire, volcano lava. They are all about 700 °C.

The fact that the wavelength of the brightest colour of black-body radiation depends on temperature means that we can tell the temperature of an object by looking at its spectrum. This allows us to determine the surface temperature of the surface of the sun (about 5600°C) and other stars. This is handy because sending a spacecraft to the sun with a thermometer to dip into it would not be feasible given the laws of science.

Black-body radiation and emission lines are what cause the colours in a candle flame. The blue light at the bottom of a candle flame is one pure colour of an emission line and the yellow light of the body of the flame is partly black-body radiation and partly plasma. Other examples of plasmas are an electric spark and the surface of the sun. The nature of plasmas will be described in *Atom part 3*.

In *Chapter 9 - Light part 2* we will describe how some people wanted to know why the black-body curve had that shape because they thought this knowledge could help them make money. They got their answer, but it was not what they wanted to hear. The scientific community was not happy either - at *first*.

How important are these emission and absorption lines? They are a bit like mother nature giving clues about atoms in Morse code. In 1868, French and British astronomers Pierre Janssen and Norman Lockyer working separately found a pattern of lines in the spectrum of the sun that did not correspond to any known element. They determined it was a new element which was then named helium after Helios the Greek god of the sun. Ref *Chapter 6 - Atom part 1*. Some years after Janssen and Lockyer had discovered helium in the sun it was found in cavities in rocks on earth. Today we use it to lift airships and balloons into the sky and to make our voices squeaky.

In *Chapter 10 - Atom part 2* discoveries will be described which describe how spectral lines form, why each element has its own pattern and what they tell us about atoms. And in *Chapter 14 - Cosmology part 2* we will see how they are utilised to find out the motion and content of extremely distant objects.

Light is made of what?

In *Chapter 2 - Charge in Solids* we saw how the discoveries of Ørsted, Ampere and Faraday had shown that an electric field could induce a magnetic field and vice versa.

At this point there was no precise way to describe these fields. An indication of what a magnetic field was like could be produced by sprinkling iron filings on paper and putting a magnet underneath. This allowed you to see the shape of the magnetic field and the direction at each point but did not say much about the strength of the field at each point. This was something engineers wanted to know so they could design more efficient machines.

In the middle of the 19th century Scottish mathematician James Clerk Maxwell came to their rescue when he found a mathematical way to precisely describe how a fluid flows. This method describes the strength and direction of the push and pull on each particle in a fluid as it expands as it moves into a larger space or as it contracts as it moves into a smaller space. It also describes the forces pushing and pulling the particles when the fluid twists or goes around a bend, a kind of three-dimensional calculus.

When he applied his mathematics to electric and magnetic fields it described the direction and amount of force felt by a small magnetic or charged object in the field. From this it predicted the effects that Ørsted, Ampere and Faraday had discovered. Furthermore, it described how the fields can change shape, bend and twist over time.

In 1865 James Clerk Maxwell published a comprehensive mathematical paper on electricity and magnetism showing how the equations predict:
- The strength and direction of all points of an electric field for any distribution of charge.
- The strength and direction of all points of a magnetic field for any magnet.
- The strength and direction of all points of an electric field created by a changing magnetic field
- The strength and direction of all points of a magnetic field around a wire carrying an electric current
- How these fields interact with each other and with time
- How charge distributes itself in a conductor of any shape

One of the most important things these equations showed was that if you have a moving charge it will create a changing electric field and this will induce a changing magnetic field. Also, a changing magnetic field will create a changing electric field. This in turn will create a changing magnetic field.

You can see where this is going. The two types of field keep supporting each other and spreading out. If a charged object oscillates, continuous oscillating and overlapping electric and magnetic fields that are spreading out will be produced. This is now known as an electromagnetic wave.

Maxwell's equations predict that a moving electric charge, or a charge that is changing in intensity will create an electromagnetic wave which spreads out from the moving or changing electric charge.

Earlier we saw that the work of Huygens et al. indicated that light is a waveform and that it should have a finite speed, but it didn't say what that speed would be. Maxwell's equations predicted a waveform, how it would propagate and, its speed. Amazingly that speed matches the best measurements of the speed of light (*Chapter 4 - Cosmology part 1*).

So, it turns out that Maxwell's electromagnetic wave was light. The stuff that caused eyes to evolve so predators could find prey. They weren't expecting that.

Another thing Maxwell's equations tell us is that the wavelengths of these electromagnetic waves can have any value. So, the infrared and ultraviolet that had been found at either end of the visible spectrum must also be electromagnetic waves. The spectrum of light (Infrared, Red, Orange, Yellow, Green, Blue, Indigo, Violet, Ultraviolet) became known as the electromagnetic spectrum. The transfer of energy via electromagnetic waves is also known as electromagnetic radiation. Other forms of radiation are encountered in *Chapter 8 – Charge in Fluids, Chapter 9 - Light part 2 and Chapter 10 – Atom part 2*.

As Maxwell's equations say that electromagnetic waves can have any wavelength experimentalists started looking for radiation with a wavelength longer than infrared or shorter than ultraviolet. Nothing was found until radio waves were discovered 21 years later as we will see in *Chapter 9 - Light part 2*.

The predictions made by Maxwell's equations do not stop there. They say that if a moving charged particle accelerates or changes its direction it must emit electromagnetic waves. This tells us why hot things emit black-body radiation. It is because as atoms and molecules vibrate they are constantly changing direction. Each time they do that, a little bit of electromagnetic radiation is emitted. Also, each time a bit of electromagnetic radiation is emitted a little bit of energy is carried away, so the particle loses a bit of kinetic energy. This means that the particle starts to move a bit more slowly and the object becomes a bit cooler. This is how things cool down as they radiate heat mostly in the form of infrared radiation.

So, with his equations Maxwell showed that electricity and magnetism are just different manifestations of the same force: the electromagnetic force. And they are the substance of light. This seems to answer the question about whether light is a waveform or comprises particles but if you are surprised that Newton had made a mistake wait for more information about this in *Chapter 9 - Light part 2*.

Maxwell's equations are a cornerstone of today's science and made modern technology possible. But they were not finished there. When Albert Einstein was working as a patent clerk he realised Maxwell's equations were telling us that light always travels at the same speed no matter the speed of the observer and emitter of the light. He drew many astounding conclusions from this, one of them being that energy must have an associated tiny bit of mass. This topic would require another book to explain it but one we will touch on in *Chapter 12 - Nucleus*.

Well done James Clerk Maxwell. If Nobel prizes had been awarded at this time you would surely have had one.

Light: animals use it to know where food and predators are. Man's observations about light went from blurred shadows to realising it is a waveform made from electricity and magnetism. Light was also to provide detailed information on the universe and structure of atoms as will be seen in later chapters. We will return to this part of the story after some chapters introducing atoms and the way charge moves through fluids.

Key points of this chapter
- Light normally travels in straight lines but is bent by the edges of objects
- Light usually appears to be a waveform but sometimes appears to comprise particles
- In nature, light comprises a mix of different wavelengths and the distribution of these wavelengths is known as its spectrum
- Light emitted by hot bodies have a particular type of distribution of wavelengths known as a black-body curve
- The spectrum of light shining through a gas has dark lines because certain wavelengths are absorbed by the gas (absorption spectrum)
- The spectrum of light emitted by a very hot gas has bright lines at certain wavelengths (emission spectrum)
- Each gas has a unique pattern of absorption and emission spectra which match each other
- Maxwell's equations show light comprises magnetic and electric fields
- Maxwell's equations predict the speed of light
- Maxwell's equations predict electromagnetic radiation can have any wavelength

4 COSMOLOGY PART 1

This chapter outlines how mankind found out how far away the points of light in the night sky were, a way for navigators to find where in the world they are and how this led to the first measurement of the speed of light.

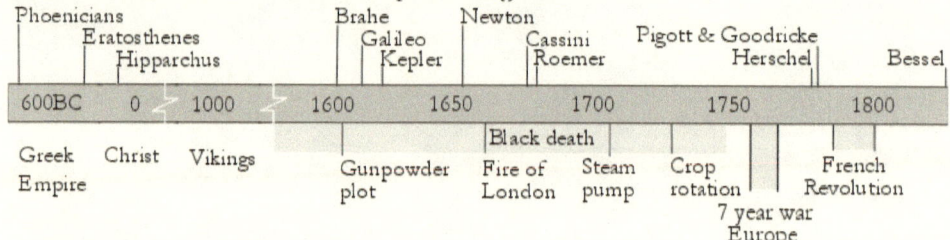

Planets are a long way away

Since ancient times people have wondered how far away the stars and planets are. But how on earth can we measure the distance? You may have ridden in a car at night and wondered why the moon appears to stay in the same place while closer things, like trees, rush past. This is called the parallax effect, which is that the angle between you and a distant object does not change much when you move from one place to another whereas the angle to a nearby object does. The parallax effect is what first enabled astronomers to measure the distance to planets and stars.

So how does it work? What you do is to look at an object from two different viewpoints and measure the distance between them. This distance is known as the baseline. Then measure the angles between the baseline and the distant object. With this information and a few sums (using trigonometry) it is easy to calculate the distance to the distant object.

In Fig 4-1 a car moves past a tree and a mountain. If you were to measure the angle between the side of the car and the mountain in position 1 then again at position 2 and read the difference in milometer readings to give the baseline you would have enough information to work out the distance to the mountain.

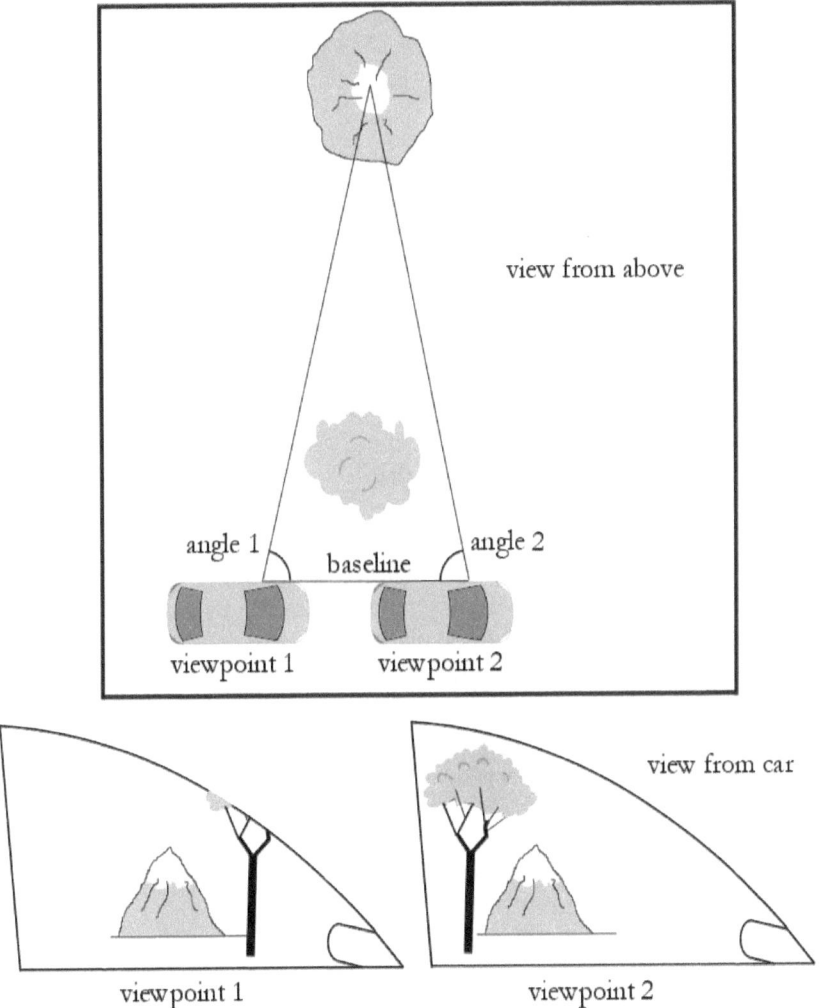

Fig 4-1 Parallax makes distant things appear to move more slowly

Your brain makes a calculation like this when you look at a beer glass, but this time the two viewpoints are your two eyes. The brain knows the distance between your eyes and the two angles to the glass from images from your eyes and works out how far you need to reach in order to grab the glass. This works fine for things up to 100 metres i.e. for things you want to chase (food) or escape from (things that think you are food). To make this work for things further away than 100 metres you would need eyes that were further apart.

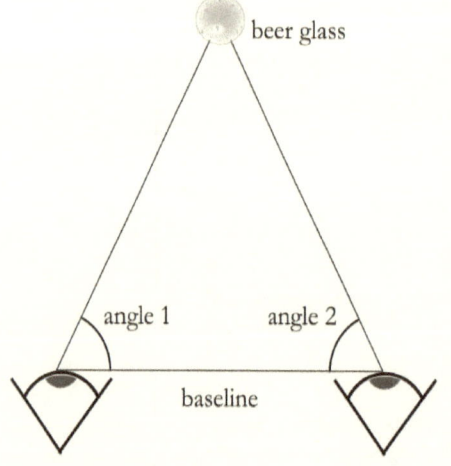

In order to measure the distance to targets that are more than 100 metres away, marksmen use a device called a rangefinder. It comprises two periscopes on their sides. This allows them to measure the angle to the target from positions 1 metre or so apart so they can determine distances much greater than 100 metres. The bigger the baseline the longer the distance you can measure.

When people first tried to use parallax to measure the distance to objects in the night sky, they could see no difference in the angles they measured to night sky objects so they realised they were going to need a bigger baseline. The biggest baseline available to us is the planet earth. So first they would need to measure that.

This measurement was provided in 200 BC when Greek mathematician Eratosthenes used the following pieces of information:
- Light rays from the distant sun are very nearly parallel. They could tell this because shadows of columns are parallel. Also, we now know the distance to the sun is thousands of times greater than earth's diameter, so light coming to us must be parallel.
- Eratosthenes knew a place where, on a particular date, the sun shone down a well with no shadow at noon.
- He knew the distance to a distant town which had been measured by counting footsteps (he would have liked a Fitbit).
- He found the angle the sun made to a vertical stick on the above date (7.2 °).

There are 50 lots of $7.2°$ in a full circle so by multiplying the distance between the two towns by 50 he got the circumference of the earth. His result was surprisingly close to the currently accepted distance of 40,000 km.

From this it was easy to work out the radius and diameter using pi.

The first person to make use of this was Greek mathematician Hipparchus who, in about 100BC, used parallax to find the distance to the moon. He used a combination of techniques because of the difficulty in measuring angles with the equipment available at that time. But in effect what he did was to find angles to the moon and use the radius of the Earth as the baseline. He used two observers, one was in a place where the moon appeared on the horizon and the other in a place where the moon was directly overhead at the same date and time. He managed to get quite close to the currently accepted figure of 382,000km.

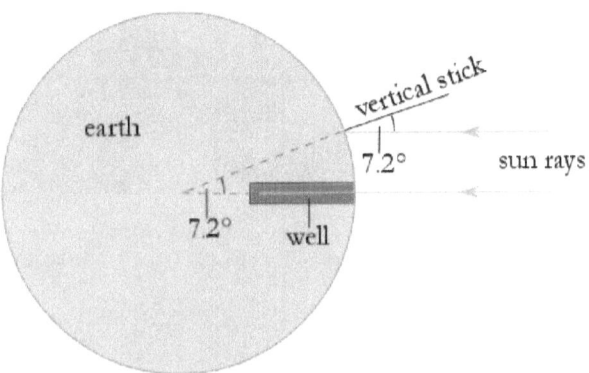

Another person who *could* have made use of this information was Christopher Columbus in 1542. It would have told him he hadn't sailed nearly far enough to reach India by sailing west from Spain. Did he keep quiet about this to please his Spanish sponsors? If a way to measure longitude had been available there would have been no doubt as we will see in a few pages.

Since before written history, people looking at the night sky have seen a vast fixed pattern of points of light rolling majestically across the sky. They noticed that most stars return to the same place on the same date each year. But some of them wander - not all over the sky but they keep within a band about halfway up the sky (for people in the same latitude as Central Europe). The moon and sun also roam in this band, so it has always been important to those interested in the night sky. The constellations in this band are a convenient way to say where these moving objects are and have become known as zodiac constellations. We now know the planets reside in a disc around the sun and when you look at this band in the sky you are looking along the edge of that disc.

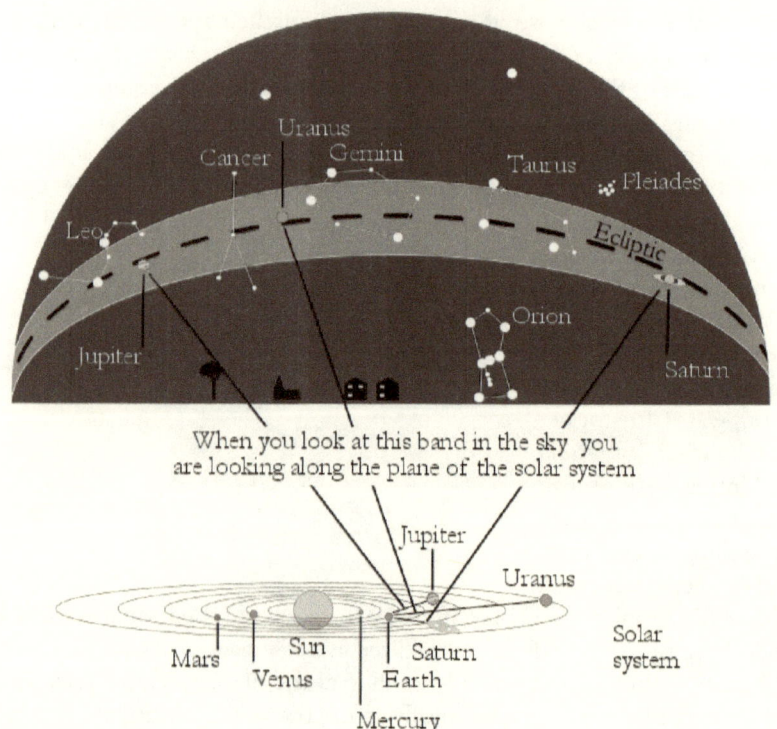

Fig 4-2 Where to find the solar system in the night sky (northern hemisphere facing south)

While observing the night sky over the millennia thousands of people have made thousands of records of what they have seen. By noticing which objects passed in front of other objects ancient people worked out that the moon must be the closest to earth followed by Mars, Venus, Mercury, the sun, then the other planets and finally the stars. But this gave them no inkling about how far away they were.

The distance to other objects remained unknown and the natural but mistaken belief that all objects rotate around the earth was a hindrance. This was until 1610 when Italian scientist Galileo Galilei produced telescopes more powerful than any that previously existed and pointed them at objects in the night sky. He discovered four moons going around Jupiter and in doing so he showed that not everything in the universe revolves around the Earth. This gave impetus to the idea that the planets revolve around the sun.

In an attempt to better understand how the solar system (the sun and planets) is organised, Danish astronomer Tycho Brahe collected a vast amount of information about the positions of planets over the years. He produced some insights, but it was German mathematician Johannes Kepler who got the most out of this data. Kepler found patterns in this data and from them he produced, and published in 1619, 3 laws which describe the shapes of paths the planets follow, how their speeds vary with their positions in their orbits and the relative distances of their orbits to the sun.

From these laws, and with the idea that planets revolve around the sun, it was possible to work out the relative distances from the sun to the planets, but no actual values. The earth-sun distance is now known as the astronomical unit (AU) and in these terms the distance between the sun and planets are[6]:

Planet	Distance from the sun
Mercury	0.38 AU
Venus	0.723 AU
Earth	1.000 AU
Mars	1.524 AU
Jupiter	5.204 AU
Saturn	9.582 AU

Fig 4-3 Relative distances from the sun to the planets

By 1672 the performance of telescopes had improved enough for Italian mathematician Giovanni Cassini to measure the distance to Mars by parallax. He sent his assistant Richter to French Guiana while he remained in Paris giving a baseline of 7000 km. This is not much larger than the radius of the Earth used by Hipparchus but the improved angle measurement that telescopes gave him allowed him to achieve the feat.

When Richter returned, they compared angles and determined the distance to Mars to be 55 million km when it is closest to earth. Using Kepler's results this gave us the distances between all of the planets and their distance to the sun. Now they knew how big an AU was - 93 million miles (150 million km). Planets are far more distant than anyone imagined.

Up to the 18th century the next planet out from Saturn had been recorded several times but it was not recognised as a planet. When German-British astronomer William Herschel (the man who discovered infrared radiation – *Chapter 3 - Light part 1*) saw it in 1781 he tracked its motion and saw that it orbited the sun and named it Uranus after a Greek god whose father was Saturn. Over the years astronomers noticed that the orbit of Uranus did not exactly match the predictions of the equations of gravity and motion that Newton had developed in the previous century (*Chapter 1 - Energy*). In 1846 French mathematician Urban Le Verrier used Newton's equations to calculate that this

---

[6] Orbits of planets are slightly elliptical but they are close enough to circles for us to ignore that fact here.

perturbation could be explained by the existence of another planet further out from Uranus and he also told astronomers where to look. Later that year German astronomer Johann Galle found it close to where Le Verrier had predicted. Le Verrier chose the name Neptune keeping to the tradition of naming planets after characters from Greek and Roman mythology.

These achievements showed that Newton's laws of gravity and motion, which work so well on earth also govern how planets move around the sun.

Light is fast

In the autumn of 1707, a fleet of 21 Royal Navy ships were returning home to Portsmouth after active service in the Spanish War of Succession in the Mediterranean. The visibility was poor, and the navigator believed they were in safe water off the coast of Brittany. In fact, they were heading towards treacherous rocky shores of the Scilly Isles. Before the mistake was realised four ships hit the rocks and about 2000 sailors were drowned, making this one of the greatest disasters in British Maritime history.

When you put a location in your satnav you can use postcode or latitude and longitude. Navigators use latitude and longitude in the same way. Latitude is a distance north or south of the equator and longitude is a distance east or west from an arbitrary north - south line going through London known as the Greenwich Meridian. It is marked by a brass strip set into the ground at Greenwich Observatory, London.

Latitude can be determined by measuring the angle of the sun above the horizon at midday and the first to do this were probably the Phoenicians in around 600BC. However, what caused the loss of life on the rocks of the Scilly Isles and many thousands before and after was that navigators could not find their longitude. In response to this disaster the British government offered a prize of £20,000 (about £3.6 million today) for anyone who could develop a method for sailors at sea to find their longitude.

The solution to this problem was to have a spin-off which remains at the centre of most scientific endeavour today.

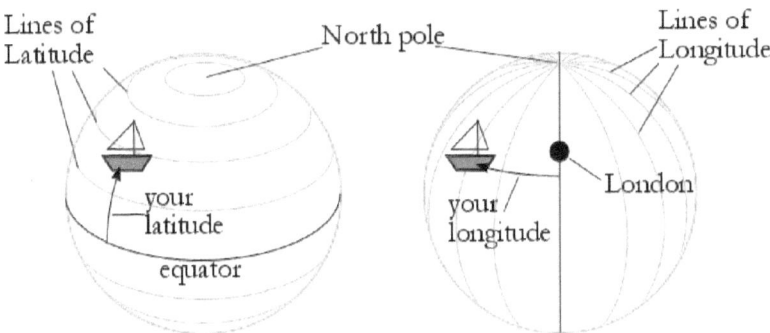

Fig 4-4 Using latitude and longitude to find your position on earth

The key to finding longitude is to know the current time in some known location and to compare it with your local time. If the sun is overhead in London and at the same time it is 60° from overhead where you are then you are on a line of longitude 60° from London. So, you can find your longitude if you have a clock set to the time in London on your ship. But unfortunately, clocks of this time relied on big pendulums swinging regularly and constantly so they needed to be kept still and at a constant temperature. On a ship at sea they were useless.

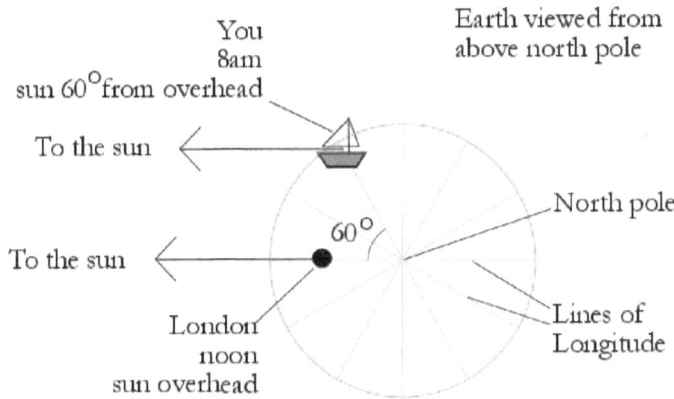

Fig 4-5 Using the difference between local time and London time to find your longitude

It was realised that if there was a clock in the sky that could be seen from anywhere on the planet that could be used to tell what the time is in London and it was proposed that the newly discovered moons of Jupiter could be used for this. This would work as long as you could see Jupiter (about 30% of the time). This was far from perfect but every little helps.

Many people thought that the moon Io which circles Jupiter once every 1.8 days could be used as a celestial clock. To use it sailors would need tables of predictions of times of Io's position relative to Jupiter. The time in London could then be established by noting Io's location and referring to the tables. And this would work no matter where you were just as long as you could see Jupiter and its moons. So, someone needed to produce these tables.

Danish astronomer Ole Rømer took on the task of building these tables and in 1676 he collected a vast amount of data on how Io's location relative to Jupiter varied with time. But he was irritated to see that the timings were not consistent. Many measurements were made over several years and the timing was found to vary by up to 22 minutes. He realised that the dates when Io was most late coincided with days when the distance between Jupiter and Earth was greatest due to their positions in their orbits. So, the apparent late arrival of Io in its orbit was due to the light from Jupiter and Io having to travel the extra distance corresponding to the diameter of the Earth's orbit. Furthermore, this information enabled the first ever calculation of the speed of light to be made. It had taken the light 22 mins to cover the 300 million km of the diameter of the Earth's orbit. This gave a speed of light of 220,000 km/s, rather short of the current figure of 299,792 km/s but this was an incredible achievement considering the accuracy and reliability of clocks available at that time. They must have thought blimey that's fast. It meant a pulse of light could circumnavigate the earth 7 times in a second.

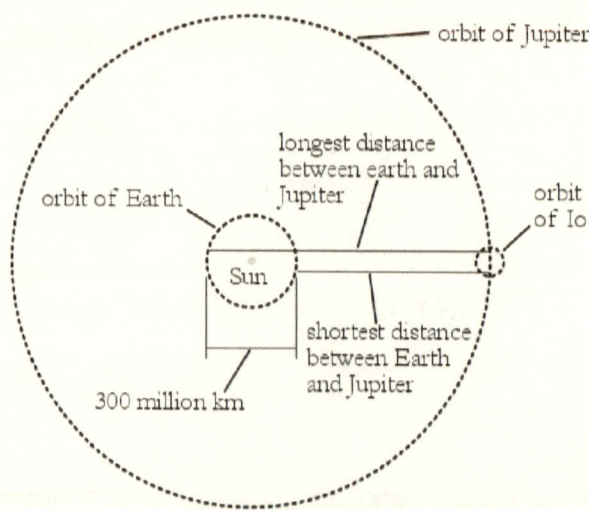

Fig 4-6 Orbits of Earth and Jupiter around the sun showing how the distance between Earth and Jupiter varies

Sailors got a method to determine longitude which would work when Jupiter was visible, and scientists got the speed of light. But what about the 70% of time that Jupiter was not visible?

The solution to this was provided by British carpenter John Harrison who developed clocks which kept good time on a ship at sea. For this he won a portion of the £20,000 prize. Various astronomical methods were used to determine time at sea until the middle of the 19th century when accurate clocks took over. (It was the discovery of one of Harrison's clocks which gave Rodney and Del Boy their fortune in a 1996 sitcom Xmas special.)

So, the quest to find a method to establish longitude gave us knowledge of the speed of light. In later chapters we will see how this knowledge led to leaps in our knowledge of the universe and how it is organised.

It was hoped the knowledge of the diameter of the earth's orbit (300 million km) could also be used as a baseline to find the distance to the closest stars if stars could be found which appeared in different places at 6-month intervals when the earth had moved half way around its orbit.

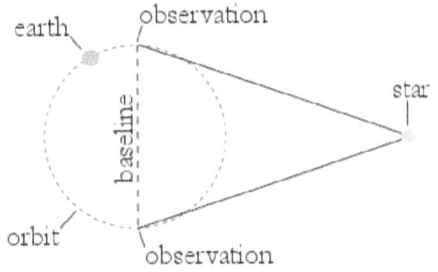

The technique used was to look at patches of sky and see if any of the stars appeared to be in a different position from where they were 6 months earlier when earth was at the opposite side of its orbit. This would mean that star was closer than the others and, from the amount that it appeared to move, the angle could be determined. This was not an easy task and at first there was no success.

Not until 1835 when Friedrich Bessel working in Konigsberg, Prussia made use of improved telescope technology to achieve this. He found that a star in the constellation of Cygnus the swan appeared to move relative to the surrounding stars every 6 months as earth went around the sun. From the tiny change in angle he calculated its distance to be 10,000,000,000,000 km. Soon, in this way, other stars were found to be similar distances.

These large numbers of zeros are very inconvenient, so astronomers came up with a much bigger unit of distance called the light year. This sounds like a measure of time but in reality it is a measure of distance, (the distance light travels in one year) and it is around 9,000,000,000,000 km. Using light years, the distance to the star in Cygnus (10,000,000,000,000 km) becomes 11 light years.

This knowledge made people realise that the stars were far too distant to consider visiting and bringing back a sample, so it was assumed that we could never know what they were made of. But as we saw in *Chapter 3 -Light part 1* spectral lines were to provide the means of knowing just that.

The use of parallax has allowed the distance to stars up to 326 light years away to be measured. Beyond that telescopes cannot measure angles accurately enough. Unfortunately, this is only a small percentage of the visible stars.

So, could we get a baseline longer than the diameter of earth's orbit to get the distances of the other stars? The only other possibility you might conceive would be a trip halfway around the (yet to be discovered) galaxy. At the end of the 20th century it was found this would take 113 million years. So, we can't have a longer baseline. Bother. To find the distance to the rest of the stars a different approach would be required.

Back in 1785 Edward Pigott announced he had found a star which varied regularly in brightness every 7 days. A few months later Dutch astronomer John Goodricke discovered one in the constellation Delta Cephei while he was working in York, England. This type of variable star is now called a Cepheid Variable. They were later to play a key role in the quest to find the distance to remote stars.

The importance of Cepheid Variables was realised in 1908 when Henrietta Swan Leavitt from Cambridge, Massachusetts catalogued thousands of Cepheid Variables in groups of stars known as Magellanic Clouds. She made the assumption that stars in a well-defined group from our viewpoint are probably close together in distance from us as well. And using this she found a simple relationship between the star's apparent brightness and the period of their variation. The longer their period the brighter they are. This means the relative brightness and period of Cepheid Variables can be used to determine their relative distance. Many people feel that this contribution deserved a Nobel prize, but this was not forthcoming.

In 1920 American astronomer Harlow Shapley, working with the world's most powerful telescope and in conjunction with other astronomers combining a variety of techniques with parallax, found the actual distance to a Cepheid Variable. Using this information, he determined the distance to many other stars. Now it was possible to determine the distance of any Cepheid Variable. From this information he drew the first map of stars in the Milky Way galaxy. Shapley's map showed the sun is not in the centre of our galaxy but in an unimportant suburb.

Galileo had moved the Earth from the centre of the universe to a suburb of the solar system and now Shapley had moved the solar system from the centre of the universe to a suburb of our galaxy. Some people were not happy.

Over the centuries, apart from finding the distance to stars, astronomers also recorded very bright stars that suddenly appeared and lasted a few months before fading away. In 185 AD, 393 AD and 1006 AD Chinese astronomers recorded such stars that were so bright they could be seen during the day. Further such sightings in 1054, 1181 and 1572 were also recorded in Europe. These sightings flew in the face of the common belief that only things close to the earth could change and the remote stars were fixed and permanent. The significance and importance of these 'temporary stars' which we now call supernovae would become apparent in the 20th century as we will also see in *Chapter 14 - Cosmology part 2*

So, parallax and Cepheid Variable stars allowed the distance to planets and the stars to be measured. Tracking of planets' orbits showed that Newton's laws work on the astronomical scale. Additionally, the quest for longitude on behalf of sailors led to the first measurement of the speed of light: information that was to form part of the bedrock of our understanding of the universe and how it works. In later chapters we will see how this information underlies the understanding of many other areas of science.

Key points of this chapter
- Observations of planets showed that Newton's laws worked for planets in their orbits around the sun
- The speed of light was found by serendipity while trying to establish an aid to navigation
- The distance to the planets and closest stars was found by parallax
- Cepheid Variables were used to find the distance to distant stars and galaxies

5 ASTROLOGY

This page is intentionally left blank.

OK, that's a bit harsh. One of the earliest benefits of studying and recording the locations of stars was that it allowed people to predict things like when the shortest day will be, when the River Nile is going to flood, when is the best time to plant and harvest crops. Much of the groundwork astrologers laid down is used by astronomers today to identify stars and their locations relative to points on earth. The trouble is that they started to believe that the location of planets had some effect over people's behaviour and characters.

Key points of this chapter
- Be careful what you believe

6 ATOM PART 1

Atoms are far too small to ever be seen with light because light does not reflect from things that are smaller than its wavelength. So, how did we come to believe they exist? This chapter covers how evidence for the existence of atoms grew until it defeated those who wanted to cling to the old ideas.

The story of the atom began around 5000 years ago when people in China took copper and tin they had dug out of the ground, melted it and mixed it together to make bronze. Apart from heralding the end of the stone age it was the first use of what we would now call an element. Later, around 1500 BC, it was found that a fairly common reddish grey rock called hematite could be heated and turned into another element, iron, making this useful element available to common people and starting the iron age.

Since ancient times use had been made of copper, sulphur, silver, iron, tin, antimony, gold, mercury and lead but the special significance of what these substances have in common and the processes used to combine them were not realised until the 19th century.

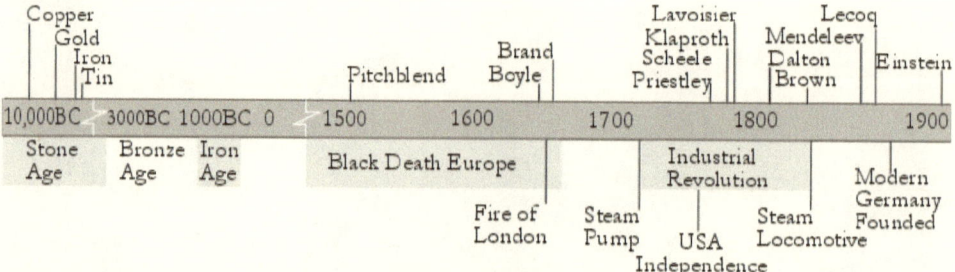

The ancient Greek atom and element

The first suggestion that the stuff of the world and life is made up of atoms came in about 440 BC when Greek philosopher Democritus proposed the first atomic theory.

His idea was that you could keep splitting any substance in two over and over again but there would come a point at which you reached particles you could not split. Those minute particles he called atoms, which was a remarkably accurate conjecture considering it was based on theory with no evidence. The Epicurean school built a philosophy on the works of Democritus; however, their ideas went out of favour when Aristotle and the Stoic philosophers came up with the idea of elements, stuff which has two properties:
- An element cannot be changed into different stuff
- All other stuff is a combination of elements.

They believed everything on earth was made up from 4 elements: earth, wind, fire and water. They rejected the idea of atoms and they thought their elements could be subdivided forever.

The first hint that earth, wind, fire and water may not be true elements came in the 2nd century BC when the Greek philosopher Philo saw droplets of water appear in a cup he held over a candle flame. Fire had turned into water. This was not supposed to happen but it was not deemed sufficient evidence to dethrone the theory of the 4 elements.

However, other areas of science progressed in 332 BC when Greek King Alexander the Great conquered Egypt and his philosophers merged these ideas with the secrets of the substances used in Egyptian embalming and with the idea of smelting. This knowledge about how some substances change when they are mixed and heated was named khem after the Egyptian word for black, referring to the rich black soil around the Nile which was so important to the Egyptian economy.

Alchemy does the groundwork

More knowledge was added in the 7th century AD when Arabs invaded Egypt and contributed their learning to the Egyptian khem. They added the Arabic prefix 'al-' to create alkhem which is thought to be the origin of the word alchemy. Alchemists developed a long-lasting secretive quest to create gold by adapting the smelting process to other cheap materials. They hoped to do for gold what smelting had done for iron. This was a quest doomed to failure because gold does not readily bond with other substances in the way iron does. Why this should be and why it means you can't make gold out of other substances will be explained later. We still get our gold by digging it out of the ground, but it is possible to create gold from other substances by a means other than chemistry and that will be covered in *Chapter 10 - Atom part 2*.

Another substance we still mine is silver. In the 1500s a silver mine in a village called Jáchymov in what is now the Czech Republic became notorious because so many of the miners became ill after working there. It was established that this illness mainly affected people who spent a long time in contact with a shiny black mineral brought up as part of the spoils of the mine. In German it was called pechblende, meaning bad luck mineral, which became pitchblende in English. It has no relation to the soft substance pitch which can be used as a sealant.

About 400 years later the importance of some of the components of pitchblende was to be discovered by a brilliant young couple in Paris. And with this knowledge the reason for the Jáchymov miners' illness became apparent, as we will see in *Chapter 12 - Nucleus*.

The Jáchymov mine had another part to play in history. Its miners were paid in coins called 'talers' from the German word Tal meaning valley. This is the origin of the word 'dollar'.

Atoms and elements are reborn

From the 1650s onwards, scientists abandoned the secrecy of the alchemist and started working in a more open manner, sharing their results, confirming or contradicting each other's findings with alternative experiments and so precipitating faster progress.

They established once and for all that earth and water *can* be turned into something else and they *are* made up from other substances, so they cannot be considered elements. Gradually they started to realise that many of the metals they knew of could not be changed into other substances and so must be elements, for example copper, iron and tin. The modern list of elements was forming.

Until this time no one had suspected the existence of different gases. Only one was known: air. One of the early investigators of elements was the British scientist Robert Boyle who poured dilute acid on iron filings and noticed that a different kind of 'air' was given off, which he collected and then found that it exploded when he exposed it to a flame. This was repeated by British scientist Henry Cavendish who established that the 'air' was an element. He also found droplets of water were left behind after the explosion just as Philo had in his cup. Subsequently French scientist Antoine Lavoisier gathered more information about these effects and gave us the name hydrogen from the Greek word for water.

Eventually it was found more convenient to use a word that had been coined by Flemish chemist Jan Baptist van Helmont: gas.

A different gas was found by Swedish and British scientists Carl Scheele and Joseph Priestley when they heated various chemicals and collected the gas given off. They realised this gas was distinctive because when a smoldering taper was put in it, it would burst into flames. Also, when Priestley breathed it, he felt more fit and alert. Furthermore, he discovered that plants produce this gas during daylight hours. The first to understand this gas was Lavoisier who performed a series of more advanced experiments showing that all combustion requires this gas to be present. It makes up 21% of the air and he gave it the name oxygen after the Greek word for sharp.

Once these and other elements had been discovered it was noticed that when they combined to form a new substance (e.g. when hydrogen and oxygen combined to form water) heat was either given off or heat was required to make it happen and the term compound was invented to mean a substance made up of two or more elements.

The compounds produced by mixing elements and adding heat were mostly seen to be very different from the constituent elements. For example: hydrogen (explosive gas) and oxygen (breathable gas) form water. Or sodium (soft metal) and chloride (poisonous gas) form the salt we use to add flavour to food.

The next thing they realised when combining elements to form compounds was that the weights of the elements used were always a simple multiple of each other. For example, when hydrogen and oxygen combined to form water the amount of oxygen used would always weigh 8 times the weight of hydrogen used. If you started with 100gm of oxygen and 10gm hydrogen 80gm of oxygen would combine with the 10gm of hydrogen to make 90gm of water with 20gm of oxygen left over. Alternatively, if you started with 1000gm of oxygen and 100gm hydrogen you would end up with 900gm of water with 200gm oxygen left over. Or, if you had too much hydrogen some of that would be left over. Always the proportions were 8 lots of oxygen to 1 lot of hydrogen. Other chemical reactions required different ratios of ingredients.

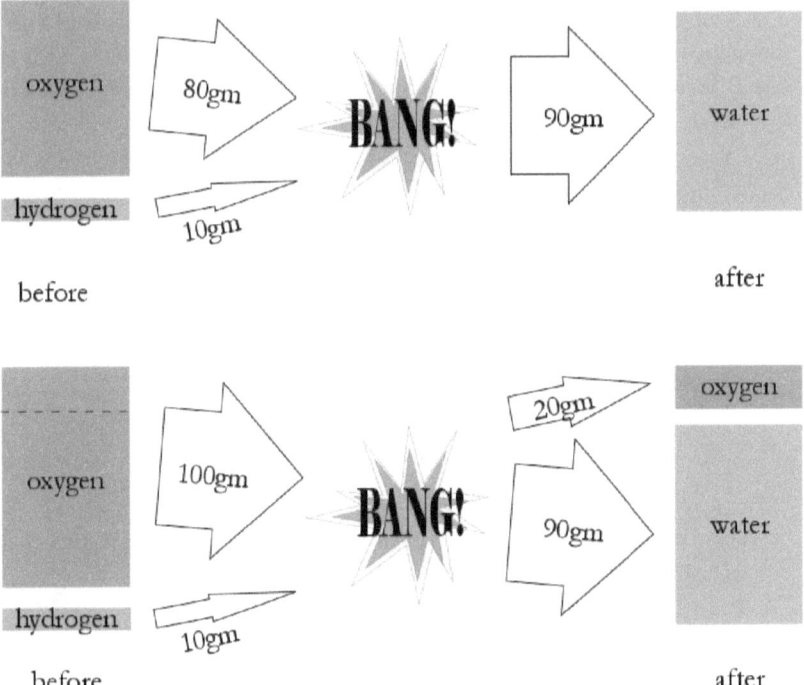

Fig 6-1 Burning with different ratios of hydrogen and oxygen

This is different from a mixture where substances combine in any ratio, e.g. salt and water.

What was found can be summarised as:
- Mixing certain combinations of elements produced a compound quite unlike the elements you started with
- Doing this sometimes meant that heat was produced OR...
- Heat was required to make it happen
- The weights of the elements used were always a simple combination

The term 'chemical reaction' was invented for this.

These findings were the result of a lot of careful experiments carried out by many scientists, most importantly: Antoine Lavoisier, Michael Faraday, Amedeo Avogadro, Marc Gaudin, Friedrich Kekule, Archibald Couper, John Dalton, Charles Wurtz, Joseph Louis Gay-Lussac, William Higgins.

What did all this mean? Some decided it was time to bring back Democritus' idea of atoms and combine it with the notion of elements. The thinking was if atoms make up elements and only certain numbers of atoms will join together when compounds are formed this would explain the simple ratios of weights of elements and the particle formed when atoms joined together was named a molecule.

In 1805 British scientist John Dalton summarised all this in a paper based on the results of many experiments by many people. It proposed:
- There is a finite number of different types of atom
- Each element is made up from atoms of one type
- All atoms of an element have the same mass
- Atoms of different elements can join together to form molecules
- All compounds are made up from molecules

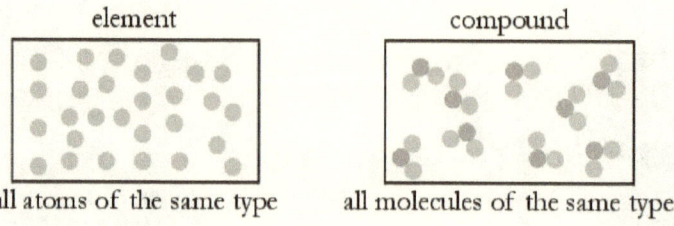

Fig 6-2 The relationship between elements, compounds, atoms and molecules

Dalton realised atoms could also explain how different gases can coexist in air. The difficulty was that in air different gases occupy the same space. With solids this can't happen, but if gases are made from molecules and atoms they could intermingle.

Another way to think about atoms, molecules, elements and compounds is to compare them with letters and words. The 26 letters used in English can be grouped together in about a million different ways to make the words of English. We now know there are 92 natural elements which can be grouped together in about 10 million different ways to make the known chemical compounds. Every year chemists find new ways to group atoms together to make new molecules to heal the sick and make new materials with new properties. And molecules have a wide range of sizes from the 3 atoms in a water molecule to 200 billion atoms in a DNA molecule.

And these atoms are the same throughout the universe. If an alien ever lands and offers to show us inside her spaceship many of the compounds we see are likely to be unknown to us and to have exotic properties but the atoms they are made from will be selected from the 92 we are familiar with.

It became clear that nearly all of the stuff around us comprises molecules. Even stuff which comprises just one element is normally made up from molecules: oxygen gas comprises molecules of two oxygen atoms and hydrogen gas comprises molecules of two hydrogen atoms. Not until the end of the 19th century were gases discovered that comprise atoms on their own instead of molecules. These are the noble gases such as helium and neon whose atoms do not form molecules but were to prove important in other ways as we will see.

As more and larger molecules were discovered writing out the element names and numbers of atoms in each molecule became inconvenient so a shorthand of using one or two letters to represent each element was adopted. H for hydrogen, O for oxygen, Cu for copper etc. In this shorthand the number of atoms of an element in a molecule are denoted by a subscript number after the element symbol. So, a molecule formed of two atoms of hydrogen became H_2 and water molecules are denoted H_2O.

From their experiments, scientists worked out the relative weights of atoms. They established, for example, that water molecules comprise 2 hydrogen atoms and 1 oxygen atom. The weight of the oxygen used when water is created from these atoms is 8 times that of the hydrogen but there are twice as many hydrogen atoms, so oxygen atoms weigh 16 times more than hydrogen atoms.

The concept of atomic mass was introduced to describe how much mass atoms of each element have[7]. It does this by showing an element's atoms mass in relation to the mass of a hydrogen atom. For example, the atomic mass of radon is 222 which means one of its atoms is 222 times heavier than one atom of hydrogen.

Some of the early atomic masses to be established are listed in Fig 6-3

---

[7] Originally the term atomic weight was used but this was later changed to atomic mass because weight differs in different gravitational fields and mass does not. So atomic mass is more appropriate. This book uses atomic mass throughout.

Element	Hydrogen (H)	Lithium (Li)	Boron (B)	Carbon (C)	Nitrogen (N)	Oxygen (O)
Atomic Mass	1	7	11	12	14	16

Fig 6-3 Atomic mass of the 6 lightest elements

At this time scientists had no idea how much mass these atoms actually had; only how many times more mass than a hydrogen atom each atom had. So, they were in a similar situation to astronomers in the time of Kepler who knew the relative distances between the planets but not their actual distance. Ref. *Chapter 4 - Cosmology part 1*.

All this is strong evidence that atoms and molecules are real and more came from the kinetic theory of gases (*Chapter 1 - Energy*). But it was still not enough to convince the majority of scientists of the 19th century who revered the teachings of Aristotle. Aristotle's idea of elements was good, but it is a little ironic that the actual elements were known of and in use while he was teaching about earth, wind, fire and water.

In most of our environment the elements are all mixed up in the form of compounds and mixtures and the same elements crop up in very different places. The iron atoms that form molecules which are good at carrying oxygen in the legs of butterflies are also used for their strength in the undercarriages of jumbo jets. And all of the elements in whatever you are looking at now were extracted from the ground or plants. And the plants extracted them from the air.

If you want to discover an element and you can't find it in its pure form, you have to extricate it from these mixtures and compounds by subjecting them to chemistry and heat not unlike the activity of the alchemists. This is a bit like a chef who wants to know what a rival puts in his trademark sauce.

The first element that was *not* discovered in its pure form was found by Hanoverian alchemist Hennig Brand in 1669. He discovered phosphorus by repeatedly boiling and cooling urine.

In 1789 German scientist Martin Klaproth got hold of some pitchblende which by this time had been found in several mines. He dissolved it in acid, mixed it with sodium hydroxide then heated the resulting yellow crystals to produce a black powder which he assumed was a new element and named it uranium after the recently discovered planet

Uranus. In fact, it was not pure. When it was purified about 50 years later it was found that uranium was a grey metal. This turned out to be the element with the heaviest naturally occurring atoms ('unnatural' elements will be covered in *Chapter 12 - Nucleus*).

The last natural element to be discovered was promethium in 1944 by scientists working on the development of the atom bomb. This time they weren't pissing about.

With all this evidence still only a minority of scientists believed in the existence of atoms and molecules, such was the grip of the ideas of Aristotle. But early in the 19th century several researchers had noticed that grains of pollen in water appeared to shift about when viewed through a microscope. The thought was that this movement must be due to some life form. However, in 1827 when British botanist Robert Brown studied this phenomenon using pollen grain that had been dead for more than a century, this too was seen to move in the same way, as did tiny chips of glass and particles of smoke. Also, if the water was warmed, the particles moved faster. This mystery became known as Brownian motion and its importance with relation to molecules will become apparent later in this chapter.

Patterns of properties create a table

As elements were discovered it was noticed that many have similar characteristics. Some, such as lithium, sodium and potassium are soft dull metals which give off gas vigorously when you drop them in water. Others, such as beryllium and magnesium, are soft silvery metals which do not react when you drop them in water. Then there are toxic gases such as fluorine and chlorine. These groups of elements became known as alkali, alkaline earth and halogen.

Group name	Characteristics
alkali	soft dull metal reacts with water
alkaline earth	soft silvery metal no reaction with water
halogen	toxic liquid or gas

Fig 6-4 Groups of elements which have similar characteristics

In the 1860s Russian scientist Dmitri Mendeleev listed what was known about the elements discovered at that time in order of atomic mass. The table in Fig 6-5 shows some of what he found.

Element	Atomic mass	Group
Hydrogen (H)	1	
Lithium (Li)	7	Alkali
Beryllium (Be)	9.4	Alkaline earth
B… O	11…16	
Fluorine (F)	19	Halogen
Sodium (Na)	23	Alkali
Magnesium (Ma)	24	Alkaline earth
Al… Na	27.4… 32	
Chlorine (Cl)	35.5	Halogen
Potassium (K)	39	Alkali
Calcium (Ca)	40	Alkaline earth

Fig 6-5 List of elements in order of atomic mass

On seeing the sequence 'halogen, alkali, alkaline earth' repeat periodically he realised it would be beneficial to arrange the elements in a table where each group formed a row, as in Fig 6-6.

H 1	Be 9.4	Mg 24	Ca 40	Alkaline earths
	B 11	Al 27.4		
	C 12	Si 28		
	N 14	P 31	As 75	
	O 16	S 32	Se 79.4	
	F 19	Cl 35.5	Br 80	Halogens
Li 7	Na 23	K 39	Rb 85.4	Alkalis

Fig 6-6 Groups of elements arranged in a table with similar elements on each line

Hydrogen does not fit in the periodic sequence, so he put it on its own on the top left corner. The rest of the top row was given over to the alkaline earths. The halogens went in the second from bottom row with alkalis taking the bottom row.

The table looked good but there were a couple of gaps between calcium and arsenic. Mendeleev made the prediction that an element should exist which would fill these gaps. To fill the gap to the right of aluminium he predicted an element should exist with properties similar to aluminium but with atomic mass between 40 and 75. Many people started looking but nothing was found until 1875 when French scientist Paul Lecoq produced an improved version of Fraunhofer's spectroscope (*Chapter 3 - Light part 1*) and used it to examine the vapour of zinc ore. He found an emission line in the violet part of the spectrum which did not match any known element. Further investigation found it had an atomic mass of 70, nicely inside the range predicted by Mendeleev. Lecoq named the new element gallium which turned out to be a soft white metal similar to aluminium. Today it is a vital ingredient of red LEDs and solar panels.

Following Lecoq's discovery, Mendeleev's table gained worldwide respect and intense interest in these gaps grew as others joined the race to find the elements which would plug them.

Near the end of the 18th century Scottish scientist William Ramsay discovered the element argon, which is a gas which does not react with any other element. He went on to discover other gases with this property. Among them was helium which he found emanating from certain rocks. Previously this had only been known to exist in the sun (*Chapter 3 - Light part 1*). In *Chapter 10 - Atom part 2* we will see how it got there. A further non-reacting gas he discovered was neon (Ne 20) which was favoured by people making advertising signs until about 2010. This set of gases became known as noble gases because 'they is a bit snooty' and don't have truck with any other element. These noble gases required a new group to be added to Mendeleev's table.

With the noble gases slotted into the list of elements ordered by atomic mass we see the idea of periodic sequence is enhanced (halogen, noble, alkali, alkaline earth).

Element	Atomic mass	Group
Hydrogen (H)	1	
Helium (He)	4	Noble
Lithium (Li)	7	Alkali
Beryllium (Be)	9.4	Alkaline earth
B… O	11…16	
Fluorine (F)	19	Halogen
Neon (Ne)	20	Noble
Sodium (Na)	23	Alkali
Magnesium (Mg)	24	Alkaline earth
Al… Na	27.4… 32	
Chlorine (Cl)	35.5	Halogen
Argon (Ar)	40	Noble
Potassium (K)	39	Alkali
Calcium (Ca)	40	Alkaline earth

Fig 6-7 List of elements in order of atomic mass with noble gases added

As scientists started to make use of Mendeleev's table, they found it more convenient to draw it sideways and they started calling it the Periodic Table as in Fig 6-8. The new group of noble gases have atomic masses which fit neatly between the halogen and alkali elements in the table, so they got their own column on the right-hand end of the new periodic table. For example, neon (Ne 20) fits between the halogen fluorine (F19) and the alkali sodium (Na 23).

Alkali	Alkaline earth						Halogen	Noble
H 1								He 4
Li 7	Be 9.4	o t h e r s	B 11	C 12	N 14	O 16	F 19	Ne 20
Na 2 3	Mg 24		Al 27.4	Si 28	P 31	S 32	Cl 35.5	Ar 40
K 39	Ca 40		Ga 70	Ge 72.6	As 75	Se 79.4	Br 80	Kr 84

Fig 6-8 First manifestation of the periodic table of elements

You may notice these atomic masses are disappointingly irregular and some of them, beryllium (Be 9.4), aluminium (Al 27.4) are not even close to a round number. Later measurements with improved equipment and techniques showed that none of the atomic masses are round numbers (although many are close).

Element	Hydrogen (H)	Helium (He)	Lithium (Li)	Beryllium (Be)	Boron (B)	Carbon (C)	Nitrogen (N)	Oxygen (O)
Atomic Mass	1.008	4.003	6.94	9.01	10.81	12.01	14.007	15.999

Fig 6-9 Modern figures of atomic masses for the lightest 8 elements

In *Chapter 12 - Nucleus* a discovery is described which explains why the atomic masses have the values they do, including why hydrogen has an atomic mass of 1.008 rather than 1.000.

Another issue you might notice is that argon (Ar) has an atomic mass of 40 and potassium (K) has an atomic mass of 39 but comes later in the table to preserve the sequence. This looks like a fudge to make the periodic sequence work but in *Chapter 10 – Atom part 2* we will encounter a discovery which justifies putting argon and potassium this way round. That discovery will also provide a simpler, cleaner number to identify elements than atomic masses and point to further information about what atoms are like inside. The reason for this lies in *Chapter 13 - Atom part 2* which also explains the shape of the periodic table.

Any final doubt that atoms and molecules exist was killed off in 1905 when German scientist Albert Einstein produced a paper which demonstrated how the kinetic theory of gases (*Chapter 1 – Energy*) agrees with the phenomena of Brownian motion. He showed how random bombardment of pollen grains by tiny atoms will cause them to move about in ways observed. Also, the fact that pollen grains move faster and further in a warmer liquid fits with the idea that temperature is a measure of how fast molecules are moving about.

The way elements combine, and the gas laws pointed to the existence of molecules and atoms and any doubt was removed by Einstein's brilliant analysis of Brownian motion.

The relative masses of atoms and the periodic sequence of the elements' properties led Mendeleev to produce his periodic table which will always be relied on by scientists in many disciplines but which was also, as will be seen later, to act as a confirmation of theories about the structure of atoms.

Each element has its own type of atom and only certain atoms will combine to form molecules which make up compounds and some of them create explosions. But how and why do they do it? And why do only certain combinations of atoms form molecules? These questions can be addressed in part 2 after the discoveries described in the next few chapters.

Key points of this chapter
- The existence of molecules and atoms was suggested by chemical reactions and the periodic table
- Elements comprise atoms which are all the same type
- Molecules comprise multiple atoms
- Compounds comprise molecules which are all the same type
- Atomic mass of an atom is its mass relative to a hydrogen atom
- Atoms of each element have a unique atomic mass
- The periodic table was created to group together elements of similar characteristics
- The elements in the periodic table are organised in order of atomic mass
- Atomic masses are mostly close to integer numbers
- Elements cannot change into different elements in chemical reactions

7 LUMINESCENCE

Most of the light we encounter emanates from objects hotter than 1000°C but occasionally we see visible light which is produced in a different way. Every day examples include hi-vis jackets, shirts that have been washed in detergent and dials of luminous watches. These are things which glow even though they are cold. This effect, called luminescence, plays a central role in our understanding of the structure of atoms.

There are two main types of luminescence: fluorescence and phosphorescence. These both involve a substance that glows with a cold light in response to being exposed to ultraviolet light, in particular sunlight, which is rich in ultraviolet light.

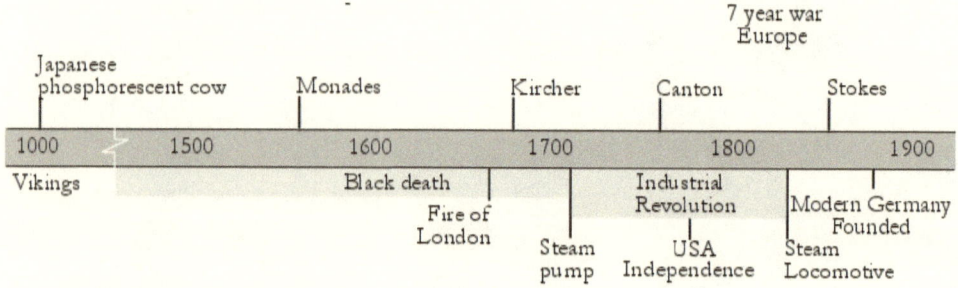

Fluorescence

One of the first written accounts of fluorescence was by Spanish medic Nicolas Monades who was testing substances in the search for new medicines. He wrote in 1565 about a liquid extracted from the narra tree (Pterocarpus indicus) which had a peculiar blue tinge when viewed in daylight which it did not have when viewed by the light of a flame.

A similar effect was seen in 1852 when Irish scientist George Stokes found that the substance which gives tonic water its flavour, quinine, appears to glow blue while it is exposed to sunlight but not if it is illuminated by the light from a flame. Using filtered light, he let one colour of light in turn fall on a test tube containing quinine in a dark room. No glow was seen when visible colours fell on the quinine but when he let ultraviolet light fall on it the quinine glowed blue.

Stokes realised the quinine was somehow changing the wavelength of the light from ultraviolet to visible - a phenomenon that became known as fluorescence. He found several materials would fluoresce, one of them being Kircher's wood extract. But in this case, instead of turning ultraviolet light to visible light, it turned blue light to yellow.

Stokes produced a law of fluorescence saying that when fluorescent materials change the wavelength of light, they always absorb short wavelength light and emit light with a longer wavelength.

A common experience of fluorescence today is items of clothing glowing under ultraviolet light in a night club. This happens because detergent manufacturers like to make white things you have washed look extra bright. So, they developed something called an optical brightener which converts ultraviolet light into white light. This causes clothes to reflect light as normal, but they also convert ultraviolet light to visible white light making them unusually bright and this makes you want to buy more of the product.

If you take a fluorescent object like a hi-vis jacket into a dark room and shine an ultraviolet torch onto walls they just look a bit violet because the ultraviolet torch gives out a bit of visible violet light. But when the ultraviolet light falls on the hi-vis jacket it glows as if it is in bright sunlight. A similar thing happens when ultraviolet light falls on a banana which is a few days old. Rings around the spots will glow.

Fig 7-1 Bananas under visible light and fluorescing under ultraviolet light

Why do offices have so many fluorescent tubes to light them? This is because a tube had been developed (based on a Geissler tube) which could produce a lot of ultraviolet light using little electrical current, so it was cheap to run as will be seen in *Chapter 8 -*

Charge in Fluids. This ability of the tube to emit intense ultraviolet light was not of much general interest until a fluorescent coating was developed similar to the optical whiteners produced for washing detergent that would turn ultraviolet light into visible white light.

So, fluorescence is also a convenient way for people to detect ultraviolet light. From the second half of the 19th century any scientist who wanted to detect some ultraviolet light would have a board painted with fluorescent paint, a fluorescent screen.

Phosphorescence

Phosphorescence is similar to fluorescence, but the glow remains for some time after the object has been removed from ultraviolet light.

The first recorded use of phosphorescence dates back 1000 years to Japanese artists who made phosphorescent paint from a preparation of shellfish shells. A Chinese document exists from this time which describes a picture of a cow in a cave in Japan which is invisible in the daytime but visible at night.

The next known use of phosphorescence was in 1763 when John Canton of Stroud, England described the preparation of phosphorescent paint by using oyster shells and sulphur.

These days phosphorescence is often used in medicine and biology where it allows intricate biological processes to be observed. Luminous watches have phosphorescent dots on their hands and numbers, so you can tell the time in the dark.

The Greenhouse Effect

The greenhouse effect in which earth's atmosphere traps heat from the sun is another example of luminescence. Visible and ultraviolet light from the sun passes through the atmosphere where it is absorbed by plants and the ground which then re-emit it as infrared light which has a longer wavelength. This is then absorbed by CO_2 in the atmosphere which steadily warms.

The greenhouse effect was so named because it was imagined that greenhouses heat up in this way. This, however, is not the case. The glass in a greenhouse blocks the ultraviolet light so it does not reach the plants. The air inside a greenhouse heats up because it is an enclosed space. If a greenhouse was made of metal, the air inside would get just as warm but blocking the light would not be good for the plants.

We have seen that the various types of luminescence are an effect of substances to turn light of one wavelength to light of a longer wavelength. In *Chapter 10 – Atom part 2* we will see an explanation of why this happens and how it gave scientists confidence in their model of the atom. The ability of luminescent material to detect ultraviolet light by turning it into visible light was to play an important role in later discoveries.

Key points of this chapter
- Some objects and liquids glow with a different colour when exposed to ultraviolet light – a phenomenon known as fluorescence
- Some objects and liquids continue to glow after the ultraviolet light has been removed – a phenomenon known as phosphorescence
- Stokes' law states that fluorescent objects absorb light of a short wavelength and emit light of a longer wavelength

8 CHARGE IN FLUIDS

Clues about the nature of the world had been found by looking at how electricity passes through metallic conductors. Could electricity flow through liquids and gases and would that tell us anything? What was found in doing this had a major impact on science and technology and is the reason the Electricity, Light and Atom chapters have more than one part.

```
Ritter              Faraday           Geissler  Crookes      Stoney     Thomson
  |                    |                  |  |                 |           |
──┼──────┬──────┬──────┼──────┬──────┬────┼──┼──────┬──────┬───┼───┬──────┼──
1800   1810   1820   1830   1840   1850   1860   1870   1880   1890   1900
         |      |                    |      |                    |      |
      Battle of Frankenstein    First Postage Big Ben         Modern   Bicycle
      Trafalgar Mary Shelley    Stamp        Tower            Germany  Motor car
                                                              Founded
```

Charge moving through liquids

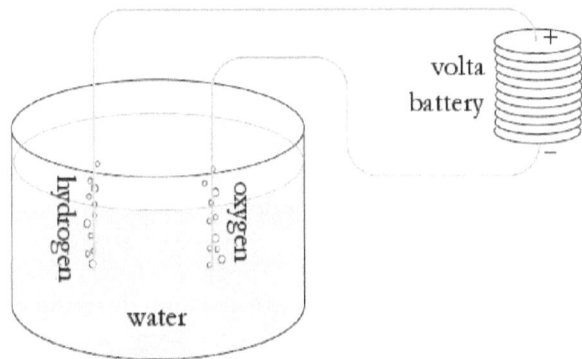

Fig 8-1 Passing current through water produces hydrogen and oxygen

After Volta had produced his battery several people experimented with passing electric current through liquids and in 1800 the man who discovered ultraviolet light (*Chapter 3 - Light part 1*), Johann Ritter, connected one of Volta's batteries to metal wires that were immersed in water. He saw gas bubbles coming off them and collected them in test tubes for analysis. Ritter found the gas produced at the positive wire was hydrogen and the negative wire produced oxygen. The water molecules had been split into their constituent atoms by a mechanism that will be explained in *Chapter 11 - Chemistry*. In further experiments he changed the type of metal in the wires and found that the metal from the positive wire had been deposited on the negative wire. He had demonstrated electroplating.

Later, in around 1830, Michael Faraday decided to find out if he could measure the electric current that flows through a liquid. He used a similar arrangement of rods of metal connected to a battery, but he included a device to measure the current. To make it easier to discuss he named these metal rods positive and negative electrodes depending on which terminal they were connected to. The positive electrode is also known as the anode and the negative one the cathode.

Faraday found that if the liquid was pure water, he could not detect any current flowing. But he did see a current flow when he dissolved various substances in the water. In particular, salt had a marked effect on the current he detected. Faraday considered this in conjunction with Ritter's finding on electroplating and concluded that whatever was carrying the charge through the liquid was different to the charge carriers he was familiar with in metallic conductors. This process causes material to leech off one electrode and deposit itself on the other, whereas when charge flows through wires no material is moved. He decided to call charge carriers which move material through liquids 'ions' after the Greek word for wanderer.

By carefully measuring the mass of the deposited material and the amount of charge that had moved, the ratio of charge to mass of these ions was measured. It was found that ions could have positive or negative charge and the ratio of charge to mass was different for each element that was transferred.

In contrast, in 1874, Irish scientist George Stoney proposed that all of the charge carriers which move through solid conductors all carry the same amount of charge. He suggested the name 'electron' for them. So, at this time scientists envisaged electrons which carry charge through solid conductors and ions which carry charge through liquids.

The technique of sending an electric current through a liquid is now labelled electrolysis and is used by scientists to separate out different substances and by manufacturers who want to coat cheap strong metals with expensive metals like gold or chrome. But for scientists the main interest was the progress in identifying charge carriers in the form of ions and electrons. We will see below how the nature of electrons was uncovered and in *Chapter 10 – Atom part 2* we will see how the nature of the ion was uncovered.

Charge moving through gases

In 1857 German scientist Heinrich Geissler set up a sealed glass tube containing pieces of metal near each end which he connected to a high voltage battery. Borrowing from the terminology of electrolysis he named these pieces of metal cathode and anode. He found that by reducing the pressure inside the tube the gas inside the tube would glow when the battery was connected. Also, he found that when the air inside the tube was replaced by certain gases such as neon a different coloured glow would appear.

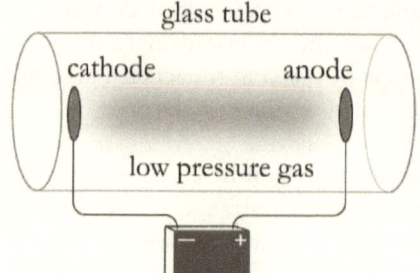

A few years later British scientist William Crookes took a Geissler tube and reduced the gas pressure inside further by using a newer type of pump. He found the glow was no longer even across the tube but contained a dark patch. When he reduced the pressure further, he saw the dark patch increase until the glow disappeared altogether. What was interesting was that the surface of the glass tube around the anode started to glow.

Glass was the ideal material for Geissler and Crookes' tubes because it is transparent, gas tight and can be formed into intricate shapes. Also, glass does not conduct electricity, so it can be used to hold electrodes. It was a stroke of luck that glass also has the property that it glows when it is hit by the substance inside a Crookes tube.

From 1869 onwards, several scientists experimented with Crookes tubes in an attempt to understand what was causing this glow on the surface of the glass. It was found that painting a fluorescent coating on the tube enhanced the effect, suggesting that the tube was creating ultraviolet light.

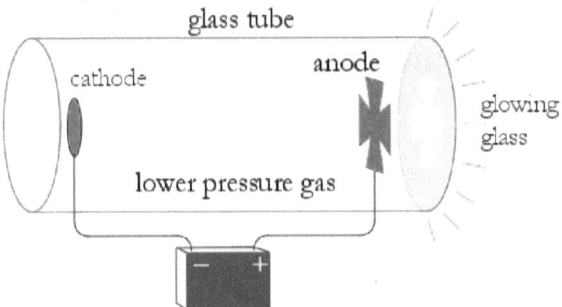

Fig 8-2 Crookes tube showing shadow of anode in the shape of Maltese cross

It was also noticed that there was a dark patch in the glow on the fluorescent coating which seemed to be a shadow of the anode. Rather like when you take a dartboard down from the wall, you see an undamaged patch where the dartboard was. This was confirmed by producing a Crookes tube with an anode in the shape of a Maltese cross. A shadow of the same shape was seen in the glow on the fluorescent coating. It seemed that something was leaving the cathode, being attracted to the anode, missing it and going on to hit the fluorescent coating. The fact that this something left the negatively charged cathode, got attracted to the anode (and missed) strongly suggested they could be tiny negatively charged particles. Because they came from the cathode, they were named cathode rays.

Crookes now put a paddle wheel in the tube at a place where it would be hit by the cathode rays. When the battery was connected the wheel turned, showing that whatever the rays were, they had some mass. Also, the paddle wheel always moved in the same direction. So, the glow was caused by something else, not ultraviolet light. He repeated these experiments using different materials for the anode and cathode and different gases in the tube but the glow on the surface of the glass remained the same.

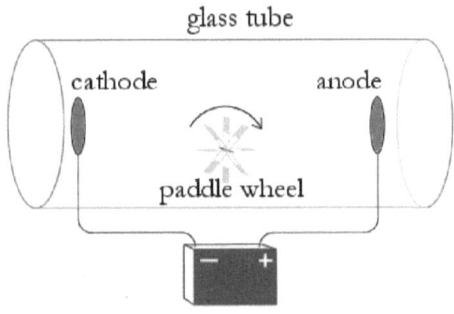

Cathode rays had several differences from the ions that Faraday had encountered in his work on electrolysis:
- They flowed in one direction only

- No deposits were made on the cathode or anode
- They were the same for all materials used in the cathode and anode

The electron is found

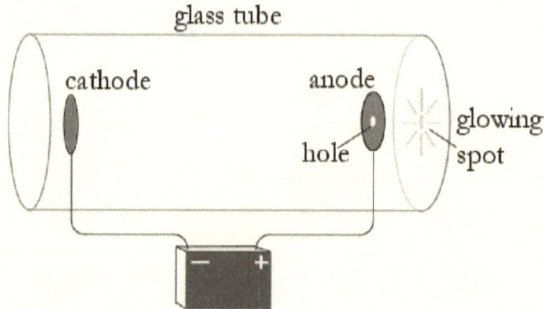

Fig 8-3 Crookes tube with hole in anode giving a glowing spot on the fluorescent coating

In 1897 British scientist JJ Thomson working at the Cavendish laboratory in Cambridge got even better pumps and reduced the air pressure in the Crookes tube still further and used an anode which had a small hole in it. This created a narrow beam of cathode rays which produced a small, well-defined bright spot on the fluorescent coating. Next, to make the movement of the spot easier to investigate, he added a large extension to his Crookes tube so the beam which caused the bright spot had to travel further from the anode to the fluorescent screen. Additionally, he arranged for the cathode rays to pass through electric and magnetic fields before they entered the extension to see what effect they had. When the strengths of these fields were varied, they both caused the spot to move.

This fact indicated the cathode rays were not ultraviolet light because no form of light is bent by electric or magnetic fields. Also, by the way the spot moved, Thomson was able to establish that the cathode rays comprise particles that possess some mass and charge. Furthermore, he measured the ratio of charge to mass for these particles. He repeated this experiment using different materials for the cathode and anode and found that the charge to mass for these particles was always the same. It became clear he had discovered the electron postulated by George Stoney: a particle present in all material.

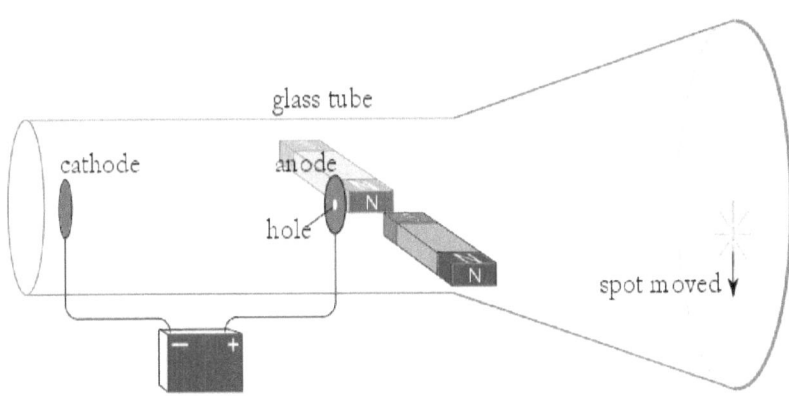

Fig 8-4 Crookes tube extended by JJ Thomson

The way the spot moved in response to the electric and magnetic fields also told Thomson what type of charge electrons have. He confirmed that it was negative.

This was rather inconvenient because by this time people had developed batteries, a telegraph system, electric motors, electric generators all of which depended on the flow of charge through electrical conductors. No one was certain, but it was convenient to assume that charge was flowing from the positive terminal of batteries and dynamos towards the negative terminal and that the type of this charge was positive. Now Thomson had shown that the type of charge which flows through conductors is negative and it flows through electrical circuits from negative terminals to positive terminals. It was far too late to update all those textbooks, so the notion of conventional current was introduced in which positive charge flows from positive to negative terminals. In reality electrons carry negative charge from negative to positive terminals.

Well done JJ Thomson, have a Nobel Prize.

At this stage, knowledge of the electron was rather like that of the distance to planets after Kepler had determined the relative distance between the planets and the sun but not the actual distance of any of them. It was also like the scientists who found the relative masses of atoms but did not know the actual masses of any of them (*Chapter 6 - Atom part 1*). In this case Thomson knew the ratio of charge to mass but not the actual value of either of them. The discovery of the actual mass of the electron is covered in *Chapter 10 - Atom part 2*.

You have probably regularly been near an object that was derived from a Geissler tube. Geissler tubes have been developed into:
- Fluorescent lighting tubes (up to 2020)
- Curly neon signs that tell us the motel is open (up to 2010)
- Cathode ray tubes used in TVs and computer screens (up to 2010)
- Radio valves (AKA vacuum tubes) (up to 1970, still used in hi fi amps favoured by aficionados)
- Geiger counter
- X-ray tube

Neon tubes are Geissler tubes which have been customised to produce a lot of light from glowing gas. Early ones contained neon gas and produced an orange-red light. Other gases were used to make 'neon' tubes of different colours.

Fluorescent tubes for lighting deserve a mention because they have been the favourite form of lighting in industry and offices for 60 years. These are Geissler tubes holding mercury vapour which produce intense ultraviolet light with only a small amount of electricity. The glass of these tubes is coated with a mixture of fluorescent materials which convert the ultraviolet light into white visible light. For a long time, they were the best way to produce light with low running costs.

The ions and electrons found by passing an electric current through liquids and gases turn out to be at the heart of why chemistry happens. In *Chapter 10 - Atom part 2* we will see an explanation of what ions are and how they form. We will also see how the study of electrons was to take science and technology into the 20th century.

Key points of this chapter
- Electrodes are pieces of metal connected to a battery and placed in a fluid
- The negative electrode is also known as the cathode
- The positive electrode is also known as the anode
- When electric current flows through a liquid, material is leeched off one electrode and is deposited on the other
- These charge carriers are known as ions which can have positive or negative charge
- The charge to mass ratio of ions depends on the material being transferred
- The electron is a negative charge carrier which is present in all elements
- The charge to mass ratio of electrons is always the same

9 LIGHT PART 2

Light. Can it have any wavelength? Is it a waveform or corpuscular? What does it do to electrons? Answers to these questions laid groundwork for new science and technologies. In this chapter we will see how forms of light at opposite ends of the electromagnetic spectrum were discovered, how Einstein and Planck resolved the disagreement about the nature of light and how this opened up the field of quantum physics.

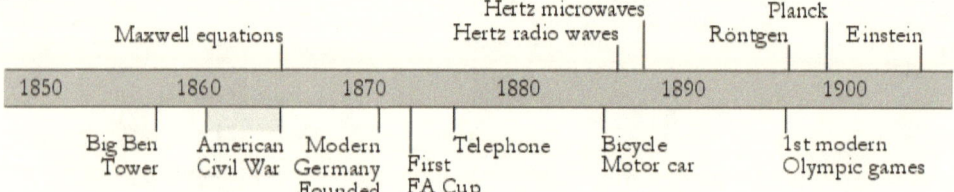

Light predicted by Maxwell is found

In 1886 German scientist Heinrich Hertz made an electrical device which produced

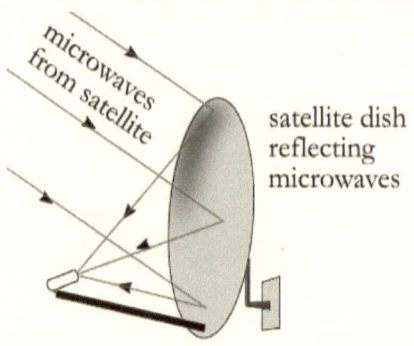

satellite dish reflecting microwaves

a fast-changing voltage across a small gap. He also produced a separate device which would make a spark jump across a gap if it experienced a voltage change at a particular frequency (a spark gap receiver). When the first device was activated the second one sparked even though there was no connection between them - it was the world's first radio, a word that was not coined until twenty years later. Hertz investigated further and found that the radiation which carried energy between the two devices travelled at the speed of light and could be reflected and refracted (bent) in the same way that light could. Soon afterwards he discovered microwaves which have a shorter wavelength than radio waves. Modern satellite dishes reflect microwaves off the dish to concentrate them on the little box that sticks out on an arm in the middle.

Radio and microwaves were clearly new forms of electromagnetic radiation whose wavelengths were much longer than any previously known. This fitted with the prediction of Maxwell's equations that electromagnetic radiation of any wavelength should exist. The unit of frequency on radios today are "Hertz" in honour of Heinrich Hertz's work. One Hertz is one cycle per second.

Radio waves and microwaves have a wavelength much longer than infrared. The electromagnetic spectrum had been extended further to:

Radio Microwaves Infrared Vis   Light **Ultraviolet**

While doing this work Hertz noticed that the spark gap receiver produced a bigger spark if light was falling on it, another link between light and electromagnetism. To investigate this further a plate on an electroscope was given negative charge, which caused the leaves in the electroscope to separate. When red light was shone on the plate the leaves remained separated indicating the plate was still charged. But when blue light was shone on the plate the leaves collapsed indicating the charge had dissipated.

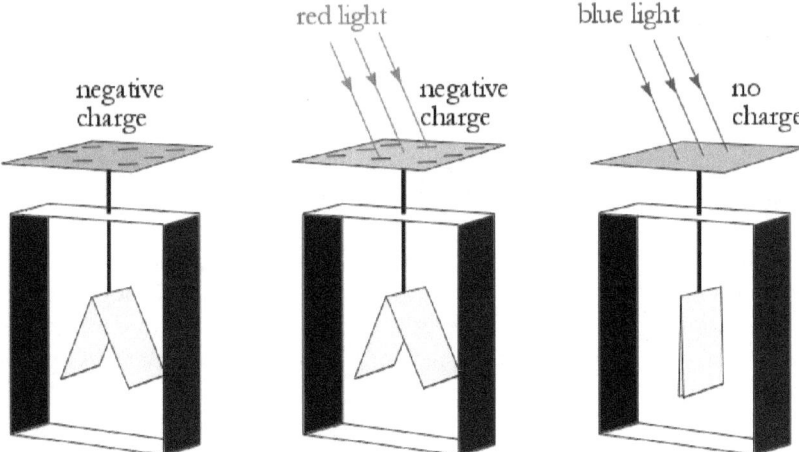

Fig 9-1 Photoelectric effect – only short wavelength light (blue) causes negative charge to dissipate

These mysterious findings became known as the photoelectric effect and were the second spark to ignite quantum physics as will be described later in this chapter.

Planck, Einstein settle Descartes-Newton dispute

Towards the end of the 19th century, German scientist Max Planck was among those who were trying to find out why the Kirchhoff's black-body radiation curve had its characteristic shape (*Chapter 3 - Light part 1*). Apart from curiosity, there was motivation from electricians who wanted to produce light bulbs that only emitted light in the visible part of the spectrum. They were miffed that of the energy they sent into their light bulbs only 2% was converted to visible light. Most of the rest was converted to infrared light and heat (which is why they feel hot). A bulb that only emitted visible light would save a lot of money and if it were cool it would last longer.

Eventually Planck produced an equation which explained the shape exactly. But neither Planck nor anyone else believed the equation described reality because if it was true, as well as the shape of the black-body curve, it was telling them three things:

- Light comes only in discrete packets of energy
- The amount of energy in each packet of energy is proportional to its wavelength
- This energy can be determined by dividing a certain number by the wavelength

This certain number is now known as Planck's constant, its importance was yet to be realised.

For them the idea that energy was delivered only in discrete packets was too outlandish to be considered real even though the maths perfectly reproduced the shape of the black-body radiation curve.

For a few years this idea languished until German scientist Albert Einstein considered these packets of energy alongside the photoelectric effect which tells us that intense red light is less effective at charging a plate than ultraviolet light with its shorter wavelength. Einstein realised the photoelectric effect happens because only the short wavelength packets have enough energy to kick electrons out of their atoms. The red ones don't have sufficient energy no matter how many of them fall on the plate. This means Planck's packets of energy are real, not just a mathematical fudge.

It is like dropping grains of talcum powder on a car roof. No damage occurs no matter how many you drop (as long as each one is blown away). But just one 1 kg rock dropped from 10 m will always dent the roof.

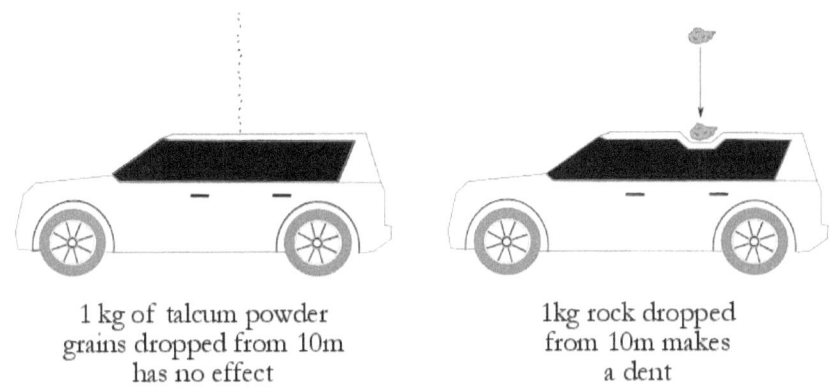

1 kg of talcum powder grains dropped from 10m has no effect

1kg rock dropped from 10m makes a dent

Fig 9-2 Effect of dropping 1kg on a car in the form of talcum powder grains and a rock

The scientific world realised it was going to have to sit up and take notice of these packets of light energy. They became known as quanta of light or photons. The sensation of sight occurs when photons enter your eye, hit the retina and kick electrons out of atoms to start a complex process which ends with your brain seeing an image of the world around you. It takes about 10 photons to produce the smallest flash that can be seen by your eye. A similar thing happens in the light-sensitive part of cameras.

Well done Albert Einstein have a Nobel Prize. Ooow... and this shows Max Planck's theory is correct, so he'd better have one as well.

The idea that light delivers energy in discrete packets called photons became known as quantisation of light. It says that red light can only deliver certain amounts of energy. And you cannot have one and a half photons' worth of red-light energy. You can have one photon's worth or two photons' worth or other multiples. Nothing in between.

Another way to look at it is that light of a particular wavelength would have energy packets only of a particular value. Each wavelength would have its own value. It is quantised in the way money in a bank is quantised. You can withdraw £22 35p, but you cannot withdraw £22 35.624p in cash. You can only withdraw multiples of 1p. A different wavelength is analogous to a different currency. In a dollar account you can only withdraw multiples of one cent.

Planck's equation tells us what the black-body curve looks like for an object of any temperature. When you heat a poker in a fire and it turns dull red, then red, then orange it does so because light comes in photons. In fact, Planck's equation tells us all things are giving off electromagnetic radiation all of the time. But for most temperatures the peak of the black-body curve will have a wavelength longer than visible light, so we won't notice it. Only when objects are heated to a temperature of about 1000°C do they start to glow red. Planck's equation did not give electricians the cool light bulb they were hoping for, but it did provide one of the key insights which led to quantum physics and, eventually, to LED lights which are cool and reliable.

Planck and Einstein had resolved the conflict between Newton and Descartes et al (*Chapter 3 - Light part 1*). They were both right. Light does come in packets and each packet is a waveform which has its own wavelength. Light has both wave and particle properties.

People had accepted that matter was quantised into atoms (*Chapter 6 – Atom part 1*), now they had to accept that light and all electromagnetic radiation is quantised into photons. The idea that photons behave both as waves and particles is now known as wave-particle duality.

Your bones are showing

Following the progress spawned by the Crookes tube many others started their own experiments. One of these was German scientist Wilhelm Röntgen who was experimenting with a Crookes tube in 1895 when he noticed that a nearby fluorescent screen glowed when he switched the Crookes tube on even though the Crookes tube was covered in black paper. He had found radiation that could penetrate paper.

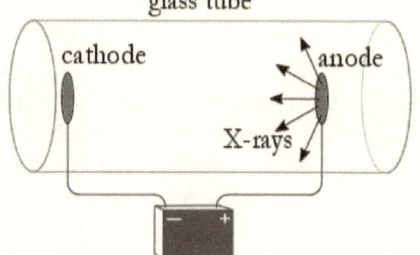

He tried putting various objects between the Crookes tube and the screen but found the fluorescent screen still glowed. Eventually he put his hand between the Crookes and the screen. He could see the outline of his bones on the screen. So, these rays penetrated his flesh but not his bones.

He knew these rays weren't infrared, visible or ultraviolet because they penetrated so many types of material. He gave these unknown rays the name X-rays (X for unknown) .

Further work showed that X-rays would darken photographic plates and he set up an arrangement with X-rays going through his wife's hand then onto a photographic plate: the first ever X-ray.

Further work showed that X-rays would darken photographic plates and he set up an arrangement with X-rays going through his wife's hand then onto a photographic plate: the first ever X-ray.

It became apparent that X-rays were emitted because electrons flew across the Crookes tube from the anode attracted by the stationary positive charge in the anode.

This decelerated the electrons to near zero speed and captured them. Röntgen knew that Maxwell's equations (*Chapter 3 - Light part 1*) say electromagnetic waves should be emitted when a charged particle is decelerated, and therefore this is what causes X-rays. The general term for radiation caused by deceleration of charged particles is bremsstrahlung which is German for braking radiation.

Well done Wilhelm Röntgen, have a Nobel prize.

In *Chapter 3 - Light part 1* we saw that black-body radiation was caused by particles changing speed or direction as predicted by Maxwell's equations. X-rays are produced in the same way, but each X-ray photon holds a lot more energy because the deceleration is more violent. Because X-ray photons have high energy Planck's equation tells us it must have a short wavelength.

When the wavelength of X-rays was measured it was found to be shorter than ultraviolet. The electromagnetic spectrum had now become:

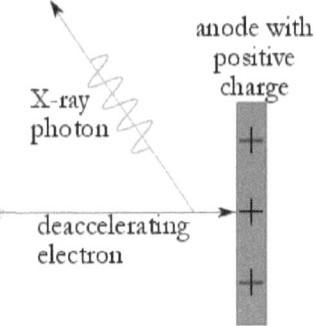

Radio Microwaves Infrared Vis   Light Ultraviolet X-rays

It becomes clear that the visible light that we use to perceive the world around us is just a small part of the electromagnetic spectrum.

We have seen that light in the form of radio waves and X-rays can have wavelengths far greater than and less than the previously known infrared, visible and ultraviolet light. So, Maxwell's prediction that electromagnetic radiation could have any wavelength was born out. Surprisingly light is a waveform and a particle at the same time, and it has the ability to deliver enough energy to electrons to kick them out of atoms. These features were to lead to new science as will be seen in later chapters.

Key points of this chapter
- Visible light is a small part of the electromagnetic spectrum
- Electromagnetic radiation comprises packets of energy called photons
- Electromagnetic radiation has both particle and wave properties
- The amount of energy in a photon is proportional to its wavelength

10 ATOM PART 2

This chapter charts how the structure of the atom was uncovered. It passes a frontier into a world that is different to the one we are familiar with - a world where it becomes increasingly difficult to use analogies with our world to explain things. But this information clears up many of the mysteries encountered in earlier chapters. Things get weirder the further you go. Hold on to your belief hats...

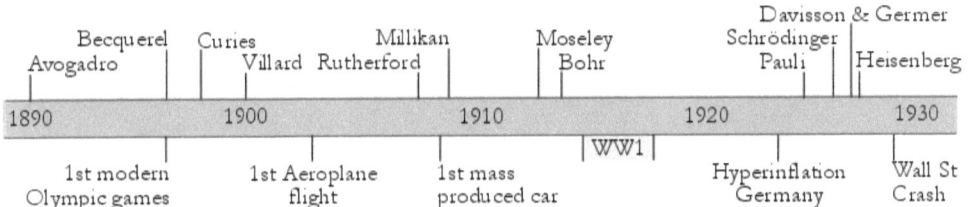

The first atom model

Normally we think of a model as something like a model aircraft. If you know the scale of the model you can measure the wingspan of the model and calculate the wingspan of the real aircraft. If you had a really good model you could see how the real aircraft looks inside and predict how it behaves. In science, models are mostly sets of equations which tell us how things like atoms look inside and predict how they behave. But these models give us access to new forms of energy, allow the creation of stronger materials, explain chemistry and much more.

The first thing people found out about atoms was how heavy they are. Scales cannot be used for this because they are not sensitive enough and they themselves are made from atoms.

The story of how the mass of atoms was found started in the first half of the 19th century when Italian Amedeo Avogadro used the kinetic theory of gases (*Chapter 1 - Energy*) to show that all gases of the same volume, temperature and pressure must contain the same number of molecules. Ah, but this did not say how many molecules. This number was first determined by French scientist Jean Baptiste Perrin, who, in 1909, came up with a number based on careful analysis of Brownian motion (*Chapter 6 - Atom part 1*). From this it was found that a gram of hydrogen has 1,670,000,000,000,000,000,000,000 atoms. So, from this they were able to establish that the mass of a hydrogen atom was 0.000,000,000,000,000,000,000,0016 gm. You're gonna need a smaller scale. But we don't need to worry about that.

In *Chapter 8 - Charge in Fluids* we saw how JJ Thomson discovered that cathode rays were electrons, but he did not show how heavy they were or how much charge they had, only the ratio between mass and charge. This is how things stood until later in 1909 when American scientist Robert Millikan carried out an experiment where he produced tiny charged oil droplets from a fine spray and determined how strong an electric field was required to stop them falling under gravity. In this way he could measure the charge of individual droplets. He found that the amount of charge on each droplet differed by the same small amount. This small amount, he reasoned, was the charge of one electron. Using this result and data from J J Thomson's work he determined the mass of the electron. It is about 2000 times lighter than a hydrogen atom which Perrin had showed weighs damn all. To this day the dimension of the electron has not been found, all we can say is that it is smaller than we can measure with our most sensitive equipment. Because electrons are smaller than atoms and a part of atoms they are designated as subatomic particles. So that's how electricity can flow through seemingly solid metal conductors: it is carried by electrons that are smaller than the atoms the conductor is made of.

Also, since atoms have no overall charge it was assumed that the atom was a blob of some positively charged substance with negatively charged electrons dotted around inside it like plums in a pudding. This became known as the plum pudding model.

Radioactivity finds the nucleus

Earlier, in the 19th century, a French grandfather, father, son team had been working together investigating phosphorescence. Their family name was Becquerel and in 1896 the son Henri was investigating the properties of uranium salts. One day he left a piece of uranium salts on top of a paper bag holding a photographic plate. When he later examined the photographic plate, he found a dark mark where the piece of uranium salts had been. So the piece of uranium salts had emitted some radiation that darkened the plate and this radiation had travelled through paper.

At that time the only radiation known that could pass through paper was Röntgen's X-rays so these became the prime suspect for the darkening. He thought these X-rays could be produced if the uranium absorbed some radiation from the sun then emitted it later as X-rays in the way that phosphorescent material absorbs UV light and emits visible light some time later (*Chapter 7 - Luminescence*).

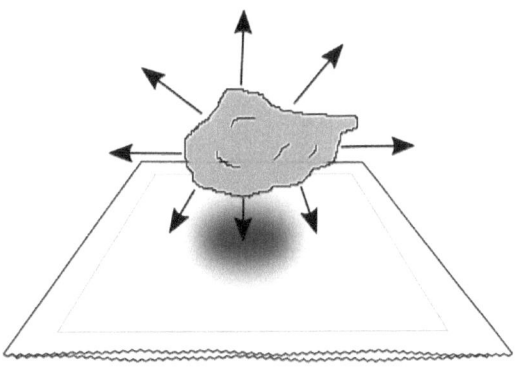

To test this, he planned to expose the substance to varying amounts of sunlight then place it on a photographic plate which was inside a paper bag. He would then be able to determine how the effect varied with the amount of sunlight the uranium had been exposed to.

On the first day of the experiment the sky in Paris was overcast. When he developed the plate, he expected to find a faint mark due to the small amount of light the substance had been exposed to. To his surprise he found a very dark mark on the plate. Later he tried the same experiment with uranium salts that had been kept in the dark for 160 hours. This too created a dark mark on the plate. The amount of sunlight the uranium had been exposed to made no difference. Not what he expected.

So, this effect was not phosphorescence because it did not require exposure to sunlight to make it happen. Also, unlike phosphorescence the effect does not quickly reduce with time. The uranium salts were fairly constantly emitting something which darkened photographic plates. We now know that the effect does reduce but takes hundreds of millions of years to fall to half of the original intensity[8]. To cause the plate to darken some radiation must have transferred energy from the uranium to the plate to break and remake chemical bonds to cause the darkening.

---

[8] This is true for the radioactivity of uranium, but other substances have a very different half-life. Some are as short as a fraction of a second.

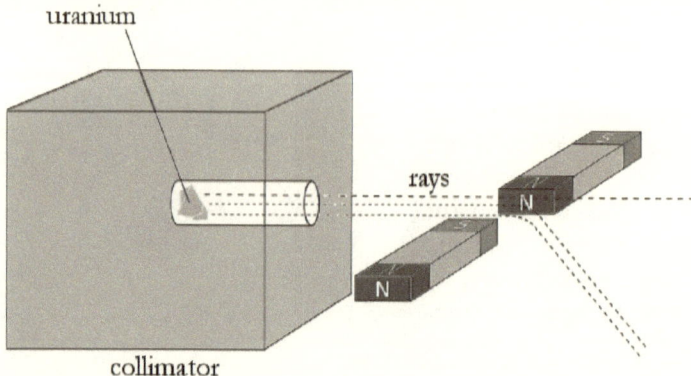

Fig 10-1 Examination of the rays emanating from uranium by passing them through a magnetic field

To investigate these rays Becquerel put some of the uranium crystals at the bottom of a hole that had been drilled in a large block of metal. Now the rays would only come out in one direction like bullets from a machine gun, an arrangement known as a collimator. He subjected the beam of rays to a magnetic field and found that most of them were bent. This told him they had charge and so could not be X-rays which have no charge.

At the time this phenomenon was not considered to be of much importance and in those unenlightened times this meant that it was possible for it to be taken up as a research project by a woman. That woman was Polish scientist Marie Curie working alongside her French husband, Pierre, in Paris. Pierre produced a method to accurately measure the intensity of this radiation and together they found that another element, thorium, emitted radiation with a similar intensity to uranium. This effect needed a name and they coined the word radioactivity.

The Curies had found that as well as uranium, the element thorium was also radioactive. They wondered if any more elements might have this property.

Marie obtained some pitchblende (the ore from which uranium is extracted) and measured the intensity of its radioactivity. They expected to find only weak radiation because it contained only a tiny amount of uranium but instead their equipment showed it was highly radioactive. This led Marie to believe pitchblende must contain another radioactive element which she set out to find. Her quest took many years of ardent work subjecting vast amounts of pitchblende to crushing, mixing with other substances and boiling to produce a tiny amount of an intensely radioactive substance which she examined using a spectroscope (*Chapter 3 - Light part 1*). The spectral lines she saw indicated it was an element that was new to science which she named radium (result). Further work produced a second element which she named polonium after her home country Poland.

So, as well as discovering two elements the Curies had shown that several elements are radioactive.

We now know all elements are to some extent radioactive (*Chapter 12 - Nucleus*) but luckily it is the rarer heavy elements which produce the most intense radiation. But as well as being a danger to life and a benefit to life in medical use, radioactivity was also to provide vital information about the structure and goings on of the atom. So, radioactivity was much more important than was at first thought.

The Curies and Henri Becquerel shared a Nobel prize for physics for this work and 7 years later Marie was to receive a second Nobel prize for chemistry in recognition of the advances she made in the isolation of elements.

Further investigation by Becquerel, French chemist Paul Villard and New Zealand scientist Ernest Rutherford showed that there are 3 distinct types of radioactivity emitted by uranium and thorium. Rutherford named them after the first 3 letters of the Greek alphabet: alpha, beta and gamma, alpha being the least penetrating and gamma the most. It was found that alpha rays are positively charged and can be stopped by a sheet of paper, beta rays have positive or negative charge and require a sheet of aluminium to stop them and gamma rays have no charge and are only stopped by several inches of lead.

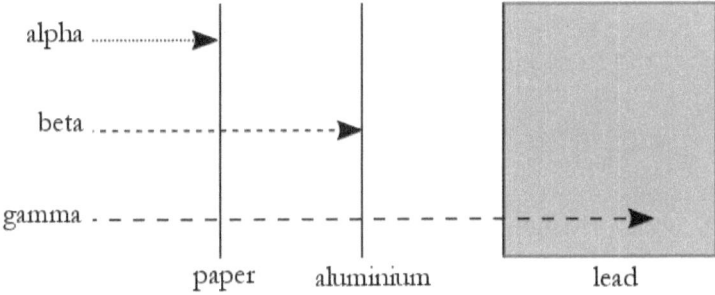

Fig 10-2 Penetration of the 3 types of radiation emanating from uranium

Just like Thomson had done for the electron, Rutherford went on to measure the ratio of charge to mass of these rays by sending them through a magnetic field and observing how much they were deflected. What he found told him that alpha rays comprise particles that are like helium atoms that have lost their electrons. Beta rays comprise particles that are almost the same mass as electrons. They come in two flavours: those with negative charge (beta-) and those with the positive charge (beta+). The plus and minus after the word beta is used to distinguish between them. The beta-particles are similar to electrons.

The only place helium is found on earth is trapped in holes in rocks near radioactive ores that emit alpha radiation. Thus, helium is created when the alpha radiation steals electrons from the surrounding rock.

It was found that gamma rays are electromagnetic rays with a wavelength shorter than X-rays. Having the shortest wavelength means gamma photons carry more energy than any other, making them the most powerful and destructive of photons.

The electromagnetic spectrum now became:

Radio Microwaves Infrared Visible Light Ultraviolet X-rays Gamma rays

So, Maxwell's prediction that electromagnetic radiation could have any wavelength (*Chapter 3 - Light part 1*) was further reinforced. From the radio waves that Hertz discovered which have a wavelength up to 1000 metres to visible light (400 to 700 nanometres (0.0000004 metres to 0.0000007 metres)) to gamma rays 10 pico metres (0.00000000001 metres).

Rutherford realised alpha rays could be used to probe the structure of atoms. To this end he produced a beam of alpha particles using a collimator like Becquerel had and directed it at gold foil which was so thin it let some alpha rays through. The idea was to see how the gold atoms deflected the alpha particles. To measure how much the alpha particles were deflected his team used a scintillator which they moved to different angles around the gold foil target. A scintillator is a sensitive fluorescent screen attached to a microscope through which a flash can be seen for each alpha particle that hits the screen. The credit for its invention goes to William Crookes, the tube man (*Chapter 8 – Charge in Fluids*).

This was the first time anyone had fired particles at other particles to determine their structure. This technique has been refined and grown over time and has allowed many important discoveries to be made. It has been likened to crashing cars in a car park, watching the way bits come flying out and using this information to determine the structure of cars. At first you might identify mirrors, glass and plastic. Only when the kinetic energy of the cars you crash in the car park is sufficient to see engines come flying out can you get an idea of what makes the car go. This is why scientists want to crash particles into others at the highest kinetic energy they can. Currently the particle smasher of highest energy is the enormous, and successful Large Hadron Collider which lies under the France/Switzerland border. (See below for what we mean by hadron.)

Rutherford asked his assistants to measure how many alpha particles came off the gold foil at each angle. As expected, if the mass and electrons were evenly distributed throughout the atom the alpha particles were deflected by a small range of angles. Finally, Rutherford asked his assistants to check if any were sent straight back to the alpha source. To everyone's great surprise a significant number of them were.

For this to happen there would have to be something heavy and dense at the centre of the atom. So, the plum pudding model of the atom had to be thrown out. The mass and electrons were not evenly distributed through the atom. Nearly all of the mass is concentrated in a tiny speck in the middle of the atom.

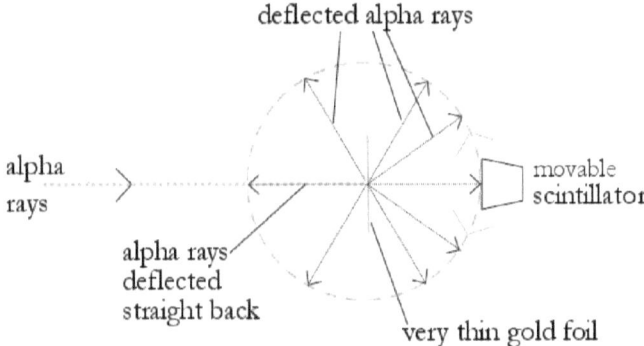

Fig 10-3 Experiment to determine the angles that alpha rays bounce off gold

This was a big shock to Rutherford. If most of the mass of the atom is confined to a tiny speck in the middle it means the atom is mostly empty space. And that means the most solid thing you can imagine, say a lump of iron, is mostly empty space. You, me and steel girders are mostly empty space. You might think when I touch another person, he or she feels solid, but it is like touching a block of foam with a giant foam hand, the hand deforms as much as the block of foam because they are made of the same stuff. When steel is subjected to immense pressure, this too will compress. The alpha ray and gold experiment gave another clue that atoms are quite unlike stuff we encounter in normal life. Later in this chapter we will look at the theory that explains why this is.

Rutherford named this speck, which is capable of making alpha particles bounce back where they came from, the nucleus. We now know the nucleus is about 100,000 times smaller than the atom in which it resides. If an atom was scaled up to the size of the Albert Hall, the nucleus in the middle would be 0.5mm (about the size of a gnat's abdomen). The popular picture of an atom is so not to scale. If the atom were 5cm wide the nucleus would be too small to see. It is hard to imagine a nucleus. As well as being so small, it is 100,000 times denser than steel. So, we are 99.99999% empty space and so are girders.

The tiny electrons orbiting around the nucleus must account for the size of the atom but make very little contribution to its mass.

Because the alpha rays have positive charge it was reasoned the nucleus must have positive charge because like charges repel each other so this is what was sending alpha particles back where they came from. Also, a positively charged nucleus balanced the negative charge of the electrons to give the neutral atom we experience. Further, if alpha particles are the same as helium atoms that have lost their electrons, this means alpha particles are in fact the nuclei of helium atoms.

So now we have a picture of an atom with a positively charged nucleus in the middle and a cloud of negatively charged electrons orbiting around it held in place by the attraction between different types of charge. However, at this stage, it was not known how many electrons and how much positive charge was in the nucleus. But now we can understand what happens when a positively charged object meets a negatively charged object. The excess of electrons from the latter rush to join the atoms of the positively charged object and they both become neutral.

However, there is a problem with this model. An orbiting electron is constantly experiencing acceleration because it is not travelling in a straight line (Newton's 1st law, *Chapter 1 - Energy*). Maxwell's equations (*Chapter 3 - Light part 1*) tell us this should mean that the electron should constantly be emitting electromagnetic radiation. This would mean it should lose energy which in turn would mean it should spiral down into the nucleus. But this is not what happens. The electrons stay in their orbits and do not constantly emit radiation. Later in this chapter we will see what is stopping it.

X-rays straighten out the periodic table

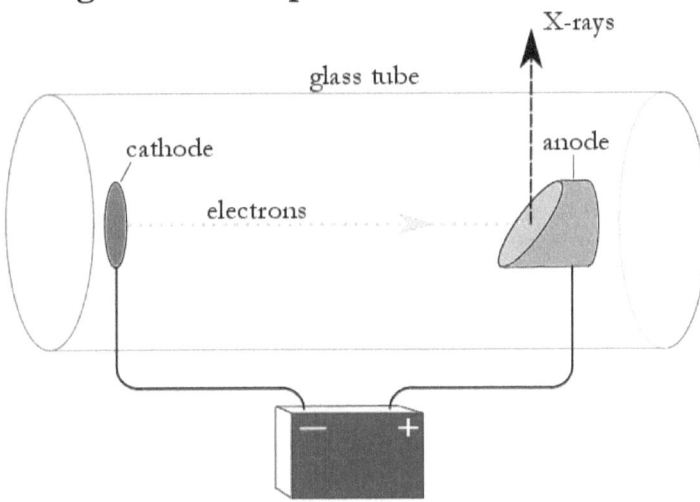

Fig 10-4 X-ray tube

In 1912 British scientist Henry Moseley, who worked with Rutherford, was investigating the production of X-rays in X-ray tubes. These tubes are based on Crookes tubes in which X-rays had been discovered but customised to produce intense X-rays. They have an anode and a cathode and electrons fly from the cathode and hit the anode which is made of a relatively massive piece of metal whose cathode facing side is angled to send lots of X-rays out of the side of the tube.

So why should X-rays be emitted from the anode? This is because as the fast electrons are captured by the positively charged anode they are slowed down to just about stationary. The conservation of energy law says that the kinetic energy has to go somewhere and, as Maxwell's equations predicted, it gets converted to electromagnetic radiation, in this case X-rays.

Rutherford made use of the X-ray diffraction work of the Braggs (*Chapter 13 - Atom part 3*) which allowed him to accurately measure the wavelength of the X-rays produced from the tube. He found the X-rays at two specific wavelengths were especially intense. These appeared to be similar to the emission lines seen in visible light discovered by Léon Foucault (*Chapter 6 - Light part 1*). Later we will see how and why these lines are created and what this tells us about the atom.

Moseley measured the wavelengths of the X-rays produced using different metals for the anode in the X-ray tube. Each metal produced emission lines at different wavelengths. When JJ Thomson carried out his experiments in which he discovered the electron (*Chapter 8 – Charge in Fluids*) he also changed the metals used for the anode and found it made no difference to the electrons that flew through the hole he had drilled in it. The phenomenon that Moseley was investigating was quite different from electrons.

Moseley created a formula which produced an integer number[9] from these X-ray emission line wavelengths based on these wavelengths.

Examples of integers that Moseley found for various elements used as anodes:
13 for an aluminium anode
14 for a silicon anode
22 for a titanium anode
26 for an iron anode
46 for a palladium anode

What was striking about this was that these numbers corresponded to the position of elements in Mendeleev's periodic table. This became known as the atomic number and atomic numbers were found to be a precise way of identifying elements (better than atomic mass). It was also better at identifying gaps in the periodic table because being integers they did not have the vagueness of the non-integer atomic masses. Moseley's atomic number became the new way to identify elements.

---

[9] An integer is a simple number that has no decimal or fractional component. Examples: 1, 2, 3. The following are not integers: 1.5, ½, 3¾.

The numbers used to identify elements could change **from**:

Element	Hydrogen (H)	Helium (He)	Lithium (Li)	Beryllium (Be)	Boron (B)	Carbon (C)	Nitrogen (N)	Oxygen (O)
Atomic Mass	1.008	4.003	6.94	9.01	10.81	12.01	14.007	15.999

Fig 10-5 Modern figures of atomic masses for the lightest 8 elements

To:

Element	Hydrogen (H)	Helium (He)	Lithium (Li)	Beryllium (Be)	Boron (B)	Carbon (C)	Nitrogen (N)	Oxygen (O)
Atomic Number	1	2	3	4	5	6	7	8

Fig 10-6 Atomic numbers of the lightest 8 elements

Traditionally, when writing about elements, if you want to remind people about the atomic number it goes before the element symbol as a subscript. So, they are now written:

Element	Hydrogen (H)	Helium (He)	Lithium (Li)	Beryllium (Be)	Boron (B)	Carbon (C)	Nitrogen (N)	Oxygen (O)
Symbol	$_1$H	$_2$He	$_3$Li	$_4$Be	$_5$B	$_6$C	$_7$N	$_8$O

Fig 10-7 Atomic numbers of the lightest 8 elements written as subscripts

We can see that the atomic masses are generally about twice the atomic number. In *Chapter 12 - Nucleus* we will see why this is and why that relation is important. Later in this chapter we will encounter the discovery which explains what the atomic number represents.

The periodic table was organised with the elements in order of their atomic mass as described in *Chapter 6 - Atom part 1*. However, there was always the concern that some lighter elements were out of order such as argon and potassium, tellurium and iodine. But when Moseley's number was used to order the elements it put them in the order of their chemical properties. Thus, Mendeleev's periodic table was further reinforced.

By the time the last natural element was discovered the periodic table held 92 elements, 2 in the first row, 8 in the second, 8 in the third, 18 in the fourth, 18 in the fifth, 32 in the sixth and 32 in the seventh row. This sequence has a regularity which seems to be telling us something. In *Chapter 11 - Chemistry* we will see what that is.

Well done Henry Moseley. Have a Nobel Prize.

How the elemental barcodes got their stripes

The next step forward came in 1913 when Danish scientist, Niels Bohr, who was funded by the Carlsberg Foundation, brought together the ideas of Planck (that light is quantised into photons and photons of a particular wavelength have a particular energy) and Einstein (that if a photon with enough energy hits an electron it will knock it out of the atom - the photoelectric effect).

He considered these questions. What happens if a photon hits an electron but does not have enough energy to knock it right out of the atom? Could this explain the spectral lines which Wollaston, Fraunhofer and Kirchhoff described? (*Chapter 3 - Light part 1*).

Furthermore, he thought if the energy of photons is quantised maybe the energy of electrons in atoms is also quantised.

Bohr proposed that electrons in an atom can only have certain levels of energy and all other levels are disallowed. He decided to use the term 'shell' to describe the physical location of electrons with a particular energy, and energy level to describe the amount of energy electrons have in a particular electron shell.

So, this raises questions: why should an electron in an atom have energy? And what does Bohr mean by energy levels?

Electrons in an atom have potential energy because they feel the electrostatic force pulling them towards the positively charged nucleus but there is something stopping it from getting there. This is like a rock at the top of a cliff which has potential energy because it feels the pull of gravity, but the cliff is stopping it from getting there. It would be natural to think that centrifugal force is what is stopping electrons falling into the nucleus, just as it stops the orbiting earth falling into the sun. But as mentioned above, Maxwell's equations say this cannot be true. Near the end of this chapter we will see what *is* keeping them from the nucleus.

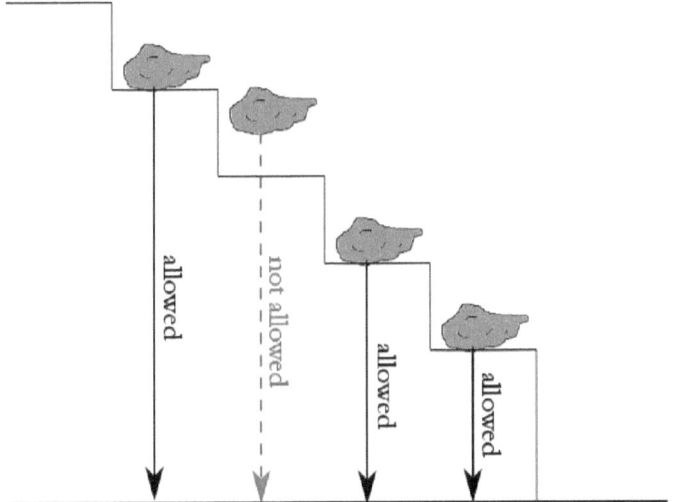

Fig 10-8 Rocks falling from a staircase as an analogy to energy levels

One way to picture these energy levels is a staircase going up the side of a cliff. A rock can be put on and kicked off any one of the steps falling down with energy corresponding to the step it was kicked off. It cannot be kicked off some arbitrary position between the steps. If a rock falls from one step to a lower one, the energy it loses will be emitted in the form of sound and vibration. This is like the rock falling off the cliff on to the car and emitting sound and heat in *Chapter 1 - Energy*. Each rock is at a different height, so the potential energy of each rock is different. We could say that each rock is at a different energy level. The rocks with the highest energy level (most potential energy) are on the step which is furthest from the ground.

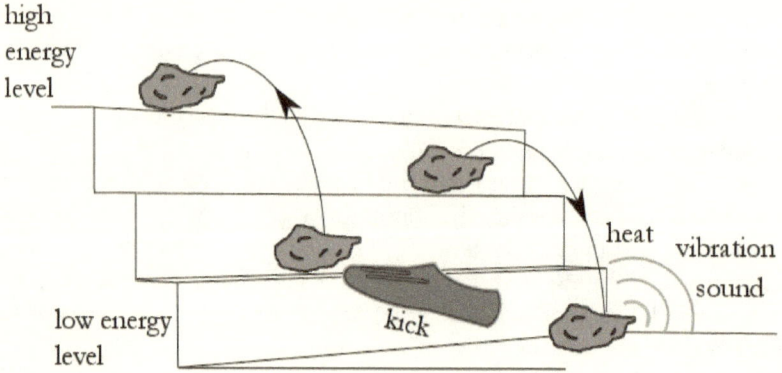

Fig 10-9 Rocks being kicked up and falling down steps as an analogy to energy levels

If a rock is moved from the first step to the top step some energy will be required, maybe in the form of a kick and that kinetic energy will get converted to the increased potential energy of the rock. If a rock falls off the second step to the ground its potential energy will be converted to heat, vibration and sound when it hits the ground or a lower step.

Bohr's electron shells are like these steps. The shell holding electrons with the highest energy is furthest from the nucleus and the shell whose electrons have the lowest potential energy is closest to the nucleus. The kick given to an electron to send it to a higher energy level would not come from someone's boot but rather, from a photon which the electron absorbs. But not just any old photon, this photon must have energy which exactly matches the energy difference between the two levels. Any other photon would be ignored.

So, this is rather different from our rock and steps analogy. If the kinetic energy given to a rock was enough to send it up two and a half steps the rock would go up, then fall back on the second step. However, if an electron were to be hit by a photon with energy sufficient to send it up two and a half energy levels it wouldn't go. The photon would be ignored, and it would go sailing on through the atom.

This is like Goldilocks' porridge which only gets eaten if it is just the right temperature. Or, another way to look at it is that the photon is only absorbed if it resonates with the gap between two energy levels. And, because photons of a particular energy all have the same wavelength this means an atom would only absorb photons of particular wavelengths. Those wavelengths are the wavelengths of the absorption lines seen in the spectra of light passing through a gas.

When an electron falls to a lower energy level it will emit a photon instead of heat vibration and sound. This photon will have energy which matches the difference in potential energy between the two levels. And the wavelengths of these photons are the wavelengths of the spectral lines emitted by hot gases.

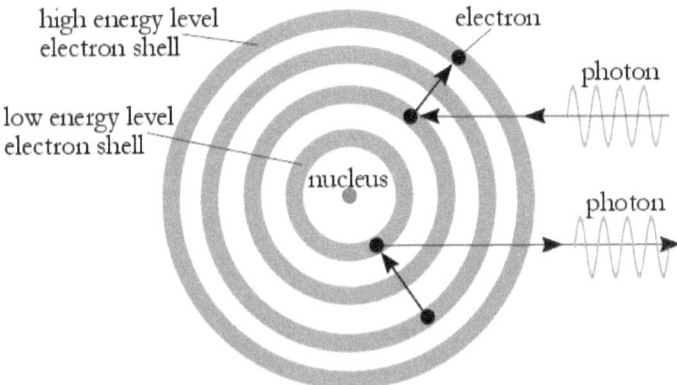

Fig 10-10 Interaction of photons and electrons in the electron shells of atoms

To calculate what the energy of each energy level would be, Bohr proposed that the angular momentum of electrons in each shell would be a simple multiple of Planck's constant[10] where that multiple is 1 for the lowest energy electron shell, 2 for the next highest and so on.

Angular momentum is closely related to the kinetic energy a thing has because it is spinning or rotating.

From the differences between the calculated energy levels Bohr was able to work out what the wavelengths of the emission and absorption lines for hydrogen gas should be. To his great satisfaction they matched those seen in experiments.

---

[10] This is a bit simplified; it is actually a multiple of $h/2\pi$ where h is Planck's constant

```
Energy levels
_____  highest
_____
_____

_____


_____


_____  lowest
```

At this point it should be mentioned that the energy levels resulting from Bohr's calculations are not all equally spaced like a staircase. The higher energy levels are closer together than the lower ones and this matches the pattern of emission lines seen in the spectrum of hydrogen gas.

Scientists represent the energy levels of an atom in a diagram comprising a horizontal line for each energy level. These are equivalent to rungs of a ladder. Each line represents an amount of energy that an electron in an atom is allowed to have.

Another aspect of electrons and energy levels is that normally an electron occupies the lowest energy level it can. If electrons are added to an atom, they will fill the lowest energy level first then, when that is full, they start filling the next higher energy level. This is because the electrons are attracted to the nucleus of the atom. This is known as the aufbau principle from the German word 'aufbau' meaning to build up. It is like beans being poured into a jar. They fill the bottom layer of the jar first and when that is full they start forming the second layer and so on. They do this because they are attracted to the bottom of the jar by the force of gravity.

So, Bohr's energy levels and the idea of photons kicking electrons to higher energy levels are real. We can now look at what they explain in the real world.

- Absorption lines in spectra
- Emission lines in spectra
- Why the above two match for a given element
- Northern lights (Aurora borealis)
- Luminescence
- Why Mosley's x-rays were well defined lines
- Positive and negative ions
- Why sparks glow

The energy levels and wavelengths of photons that match the energy differences that Bohr calculated for electrons in the hydrogen atom exactly matched the absorption lines that are seen in the spectrum of light shone through hydrogen gas.

Also, Bohr reasoned that after an electron has been kicked up into a higher energy level by a photon this will leave a place vacant for an electron in the original energy level. Sometime later the electron may fall back to that energy level. When it does so, it will emit a photon. This will have the same energy as the energy difference between the two levels and the same wavelength as the original photon. So, this explains the emission lines we saw in *Chapter 3 - Light part 1* and why their wavelengths match the absorption lines' wavelengths.

An explanation of fluorescence and phosphorescence were next on the list of successes for energy levels. A high energy UV photon (which has a short wavelength) hits an electron which absorbs it and shifts up to a higher energy level. Later it drops back to an intermediate level and emits an electron of a smaller energy (and longer wavelength) in the visible part of the spectrum.

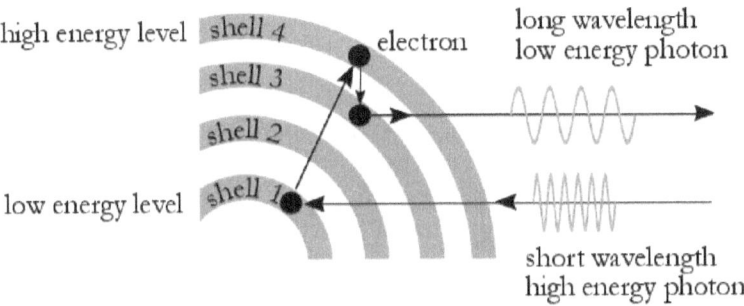

Fig 10-11 Different photons being absorbed and emitted as electrons move between different energy levels

In the case of phosphorescence, the electron remains at the higher energy level for a while before falling back down. That is why phosphorescent material gives off light for some time after the UV source has been removed. Also, it gives us an explanation of why the fluorescent coating on Crooks tube glowed and scintillators (see above) flash when a particle hits it. Electrons are being knocked out of atoms and they emit a photon when they fall back.

In arctic regions the phenomenon of 'Northern Lights' can be seen when the earth's magnetic field pulls charged particles that have been emitted by the sun into the arctic atmosphere. These fast-moving charged particles cause electrons in the atoms of gases of the atmosphere to be moved to a higher energy level. As each electron falls back to its original energy level it emits a photon whose energy is equivalent to the difference in energy between the two levels. The most common colour seen in Northern Lights is green when charged particles from the sun interact with oxygen atoms[11] in the atmosphere.

---

[11] Oxygen in the atmosphere comprises pairs of oxygen atoms in molecules. This affects the energy levels, but the way electrons move between them and emit photons is just the same.

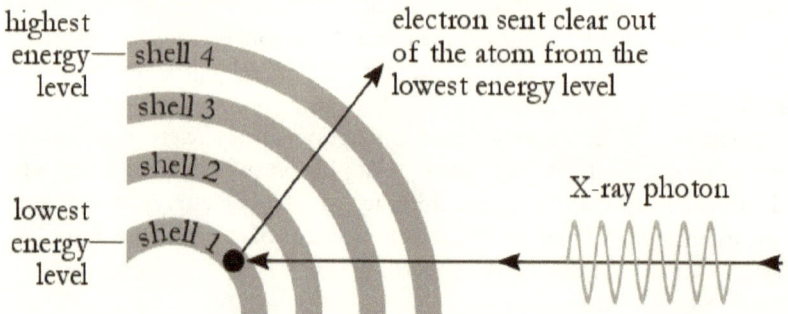

Fig 10-12 X-ray photon knocks an electron clear out of an atom from the shell with the lowest energy level

The fact that Moseley's X-rays are well defined lines and that they are so closely associated with the properties of the atoms they come from was the next thing to be explained by Bohr's energy levels. Incoming X-rays had knocked electrons clear out of the atom from the lowest energy level. To do this, they didn't need to have a special energy, it just had to be sufficient to knock the electron out of the atom from the lowest level. If it had more energy than it needed this just means the electron will fly off faster.

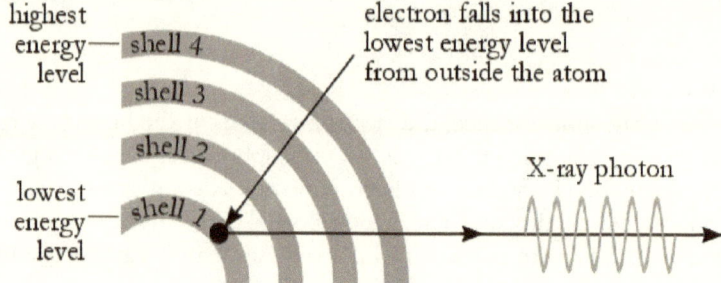

Fig 10-13 Electron falls into the bottom shell of an atom and emits an X-ray photon

Then later, another electron from outside the atom falls down to fill the vacancy created in the lowest energy level by the departure of the electron that was hit.

This incoming electron will have little or no energy because it came from inside the structure of the metal of the anode. So, the energy difference between when it was outside the atom and in the lowest energy level is the same each time it happens. So, the X-ray photon emitted in this process always has the same energy and because the energy of the photon is proportional to its wavelength it will always have the same wavelength.

To picture this another way, imagine a boy at the bottom of a well which holds stones of equal mass. He is throwing the stones out of the well. It doesn't matter how hard he throws the stones as long as they go out of the well. They will have a range of kinetic energies as they leave the well. Now he decides to tidy up by returning the stones to the well. He takes each one and drops them into the well. They all hit the bottom of the well with the same speed because they were all dropped from the same height and so they have the same kinetic energy. The bang they make as they hit the bottom of the well is the same. The only way to change the intensity of this bang is to move to a well with a different depth. Similarly, the X-ray photons emitted when the electrons fell into the atom to fill the vacated space in the lowest energy level all have the same energy, and therefore, the same wavelength. If this is done with atoms from a different element these atoms will have a different set of energy levels. So, when an electron falls to the lowest energy level of these atoms it will gain a different amount of energy.

Thus, the amount of energy an electron gains by falling into the lowest energy level depends on the element of the atom. And so, this will affect the photon that is emitted when the electron falls, and this means the wavelength of the X-ray photons emitted in this way are unique to the element. The importance of this will become apparent in the next section.

An atom which has gained extra electrons or lost one or more electrons is known as an ion, the particle first identified by Faraday as mentioned in *Chapter 8 - Charge in Fluids*. There are several ways an atom can be ionised:

- By a photon which has enough energy to knock an electron away from the nucleus which is attracting it
- By a nearby positively charged atom which pulls an electron harder than its nucleus does
- By making it so hot that when it bashes into another atom the collision energy is sufficient to remove an electron

The amount of energy required to remove an electron from an atom is known as the ionisation energy of the atom. Different atoms have different ionisation energies, and this has a big effect on an element's chemistry as we will see in *Chapter 11 - Chemistry*.

In *Chapter 1 - Energy* we saw three states of matter: solid, liquid and gas. In a solid, molecules are held in a close fixed relation to each other. In a liquid they are just about as close to each other, but they can move over each other and in a gas, they are separated and are flying about all over the place.

A plasma is a state of matter where most atoms have lost electrons. This occurs in a very hot gas where electrons have effectively been shaken off their atoms and go flying about all over the place bumping into each other. Examples are sparks, lightning bolts and flames. They glow because as the fast electrons encounter each other they change each other's speed. And Maxwell's equations tell us this will cause photons to be emitted. This is why sparks and lightning bolts glow. This is also true of flames but there are two other reasons for the glow of a flame as we saw in *Chapter 3 - Light part 1*.

In the case of the early Crookes tubes (*Chapter 8 - Charge in Fluids*) the diffuse glow that was seen was due to electrons in the gas being pulled up to higher energy levels by the passing electrons then dropping down to their original levels, emitting a photon as they did so.

This model of the atom with electrons surrounding a positively charged nucleus gives a clear picture of what is happening in charged objects. A positively charged object is one which has lost some electrons and a negatively charged object has extra electrons. When a positive object touches a negative one (as when a conductor touches the two metal parts of a Leyden jar) the excess electrons in one object are attracted to the other so they rush across to even things out. Now both objects have equal amounts of positive and negative charge and we perceive them to be neutral.

This idea of electron energy levels explains so many mysteries, is one of the key theories which makes electronics possible and is the basis of a theory which now explains how and why atoms join together to form molecules which is the heart of chemistry.

Electrons with higher energy levels orbit slightly further from the nucleus. The locations of these orbits are known as electron shells.

Well done Niels Bohr. Have a Nobel Prize.

But still the question remains: why do electrons not emit electromagnetic radiation and spiral down into the nucleus? What is holding them in the electron shells?

Rutherford splits the atom and finds the proton

Bohr's theory is very successful, but it is still not the whole story because even though it explains so much it ignores the fact that no electromagnetic radiation is emitted by electrons thought to be orbiting the nucleus and the maths that goes with it can only predict the electron energy levels for hydrogen, the simplest atom.

In 1917 Rutherford performed experiments where he fired alpha particles at nitrogen gas. He was surprised to find that oxygen and hydrogen were produced. This was surprising because the element nitrogen had changed into the element oxygen with a side order of hydrogen. But, in *Chapter 6 – Atom part 1,* it was established that elements cannot change into other elements in a chemical reaction. So, this was not a chemical reaction. Something else had happened. It was some other sort of reaction. This reaction can be written as:

Alpha particles + Nitrogen ➔ Oxygen + Hydrogen

Before we look at Rutherford's analysis of this a word about alpha particles. Earlier in this chapter, in the section *Radioactivity finds the nucleus*, it was established that alpha particles are the nuclei of helium atoms. So, the reaction could also be written as:

Helium + Nitrogen ➔ Oxygen + Hydrogen

The key which unlocked the mystery of nitrogen turning into hydrogen and oxygen when hit by alpha particles is Moseley's atomic number. Using the atomic number information in Figs. 10-5 and 10-6 above:

$_2$Helium + $_7$Nitrogen ➔ $_8$Oxygen + $_7$Hydrogen

In terms of atomic numbers alone:

2 + 7 ➔ 8 + 1

The total of the atomic numbers before and after the reaction was 9.

Rutherford concluded that the atomic number is the number of hydrogen nuclei in each atom. The hydrogen nucleus is itself a particle - the particle that carries the positive charge in atoms. He named it the proton. Nitrogen has 7 protons, Oxygen 8 and so on.

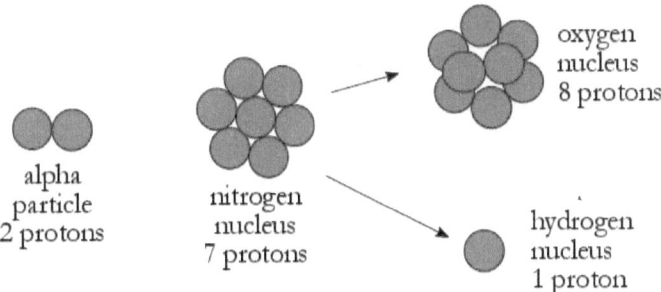

Fig 10-14 Alpha particle changing nitrogen into oxygen and hydrogen nuclei

The atomic number that Moseley had produced turned out to be the number of protons in the nucleus of every atom in an element.
Now the list of sub-atomic (smaller than an atom) particles had increased to 2, the electron and the proton.
The reaction in Rutherford's experiment was not a chemical reaction, it was the first observed nuclear reaction. Newspaper headlines labelled it "Splitting the Atom".
The dream of alchemists from hundreds of years ago had been realised (sort of). Elements can be changed into different elements by firing radiation at them. So, can we get the lead off the roof and start turning it into gold? Yes, but the trouble is you will have to spend millions of pounds on the equipment to produce a lump of gold the size of a pinhead. It is still far cheaper to go to a jewellers and buy it from them.
Further study of protons showed:
Protons have the same charge as electrons except that the charge is positive
Atoms normally have no overall charge so the number of electrons in an atom

normally matches the number of protons

The protons are packed tightly together in the nucleus but they are all of the same charge so they must be trying to fly apart, so some force must be holding them in. This force will be met in *Chapter 12 - Nucleus.*

But there was still the mystery of why atomic masses were not round numbers. Also, the masses of protons and electrons in an atom did not add up to the mass of the atom. For example, neon has 10 protons, but its atomic mass is the equivalent of 20.18 protons. This was solved a few years later as will be seen in *Chapter 12 - Nucleus.*

The proton is the first of a class of particles now called hadrons to be discovered and it the particle that the Large Hadron Collider spends much of its time crashing them into each other.

Matter waves shape atoms

In the section above (*How the elemental barcodes got their stripes*) we saw that the energy of electrons in atoms is constrained to a few allowable levels as predicted by Niels Bohr. People wondered why this should be. In 1924 French scientist Louis de Broglie proposed the theory that if a photon sometimes acts like a particle and sometimes like a wave then perhaps an electron can sometimes behave like a wave. The same maths that so successfully described the behaviour of photons of light going through Young's slits or a diffraction grating was to be applied to particles. This seems bizarre but in 1927 an experiment carried out by American scientists proved just that.

The wavelength de Broglie calculated for the electron is so small that any diffraction grating would need gaps similar in size to an atom. So too small to make. What Davisson and Germer did was to use a crystal of nickel as a diffraction grating. They fired electrons at it with a set up similar to that which JJ Thomson had used when firing electrons at a fluorescent screen. The pattern the electrons formed when they hit the screen was the same pattern that light makes when it passes through a diffraction grating.

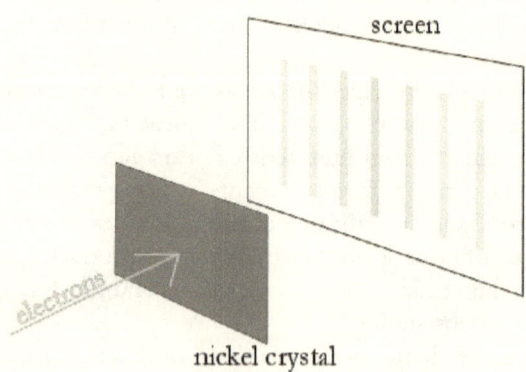

Fig 10-15 Electrons passing through a nickel crystal

So, electrons can behave like waves and they have a measurable wavelength. Also, electrons exhibit wave-particle duality in the same way that photons do (*Chapter 9 – Light part 2*). This seems very unlike how matter behaves in our world, but it led to an alternative way to predict the energy levels of electrons in an atom. Additionally, it solved the mystery of why only certain energy levels exist for electrons in an atom. To understand this, we need to take a look at harmonics.

If you pluck a guitar string in the middle of the string a pleasant note is produced. If it is plucked close to one end it sounds different because the string vibrates in a different way.

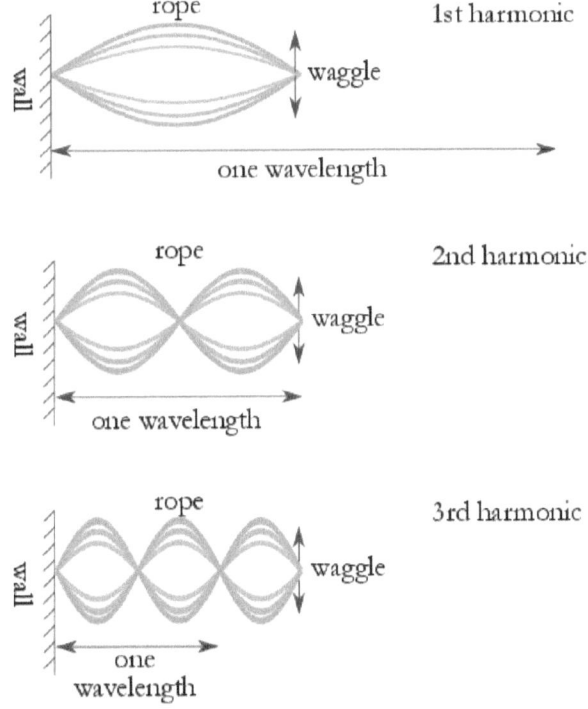

Fig 10-16 First three harmonics of a piece of rope

It is difficult to see how a guitar string vibrates but a rope with one end tied to a wall makes this easier to visualise. The free end of the rope can be waggled in such a way that the rope has a single curve as in the top diagram of Fig 10-15. If it is waggled twice as fast the rope will form two curves as in the middle diagram. And three times as fast gives the rope three curves as in the bottom diagram. These three modes of vibration are known as the first, second and third harmonics of the rope. Although parts of the rope are moving up and down the wave itself does not move unlike the wave on the surface of a pond or a sound wave. For this reason, the type of wave seen in the rope is known as a standing wave.

When the guitar string was plucked in the middle it vibrated mostly with its first harmonic. When it was plucked near the end it vibrated with a combination of first, second and third harmonics (and maybe more). It is the combination of harmonics and their relative intensities that give musical instruments their distinctive sounds.

In the middle diagram of Fig 10-15 (second harmonic) the rope was vibrating with a wavelength equal to the length of the rope. In the top diagram (first harmonic) the rope was vibrating with a wavelength equal to twice the length of the rope. And in the bottom diagram (third harmonic) the rope had a wavelength one and a half times the length of the rope.

Putting this another way for the first harmonic 1 half wavelength equals the rope length, for the second harmonic 2 half wavelengths equals the rope length and for the third harmonic 3 half wavelengths equal the wavelength. The rope can only vibrate in a way that an integer number of half wavelengths are equal to the length of the rope.

Standing waves can also be set up in a ring-shaped object like a hula hoop as in Fig 10-16.

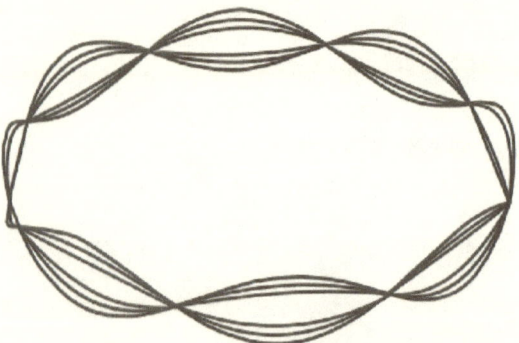

Fig 10-17 Standing wave in a ring-shaped object

Ring shaped objects can only support standing waves that are an integer number of full wavelengths. The ring shown in Fig 10-16 is vibrating with 4 full wavelengths.

The de Broglie waves associated with electrons in an atom can be pictured as a ring-shaped object vibrating with standing waves whose wavelength is equal to the de Broglie wavelength of the electron. So, electron orbits are only possible where the circumference of the orbit is 1,2,3,4… de Broglie wavelengths. These correspond to the electron shells of the atom. And so only certain radii are possible for electron orbits around an atom. Since the energy of the electron depends on the distance from the nucleus only certain energy levels are possible. And so de Broglie's theory had explained Bohr's energy levels.

Bohr's atomic theory was now on a firmer footing, but issues remained:
- Each element has its own set of spectral lines which means they have their own set of energy levels
- Bohr's theory cannot predict the energy levels of atoms other than hydrogen
- It was not understood why electrons don't spiral down into the nucleus which is attracting them
- Molecules have definite shape which effects how they behave
- The periodic table has a particular shape

In 1926 Austrian scientist Erwin Schrödinger was taking a skiing holiday in Arosa, Switzerland. He had been working with de Broglie's waves trying to answer these questions when he had a mathematical breakthrough. He produced 'wave equations' based on the de Broglie wavelength of electrons and Planck's constant. They operate in 3 dimensions and describe a 'sort of' orbit, known as an orbital, that electrons take around any atom no matter how many electrons it has. I say 'sort of' orbit because it is quite unlike the orbits that a planet takes around the sun. As mentioned above, electrons cannot orbit the nucleus in the way planets orbit the sun because this would cause them to continuously emit photons which are not seen. Orbitals are intriguing 3D shapes which represent the probability of finding an electron at any given place in the atom. Some orbitals are spherical while others are shaped like dumbbells or party balloons. These orbitals give elements their properties and control how atoms bond together to form molecules as well as the shape of those molecules. The shapes of molecules has many consequences. One is that it allows water to dissolve compounds like salt as we saw in *Chapter 8 - Charge in Fluids*.

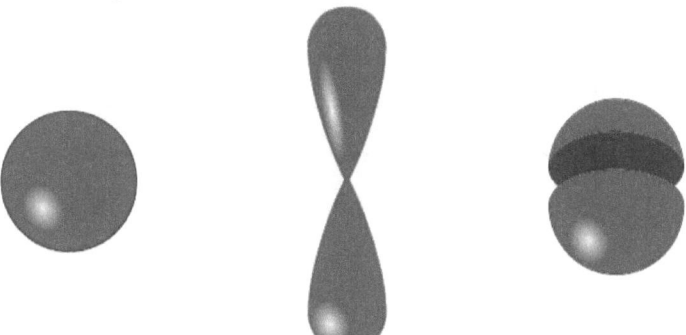

Fig 10-18 Example shapes of orbitals

Once this work was extended by Austrian scientist Wolfgang Pauli four numbers 'fell out of the bottom' of the equation. These became known as quantum numbers. It was established that each orbital had four quantum numbers associated to it a bit like four pieces of information make up the address of a house: house number, road name, town name and postcode. And just as no two houses in a country can have the same address no two electrons in an atom can have the same set of quantum numbers. This restriction is known as the Pauli Exclusion Principle.

These quantum numbers reflect the fact that all attributes are quantised at the atomic scale. German scientist Werner Heisenberg established that the first quantum number corresponds to the electron shell (or energy level) that the orbital resides in. The second and third are associated with the shape of the orbital and the magnetic orientation of the orbital and they also determine how many orbitals each energy level can have. And the fourth is the spin of the electron.

There are restrictions on the values some of these numbers can have. In particular, the electron spin can only take on one of two possible values. We can think of this as meaning one is spinning in one direction and the other is spinning in the opposite direction. This, coupled with the Pauli Exclusion Principle, means that an orbital can only hold two electrons.

When all of this was combined, a table could be produced showing the number of electrons allowed in each electron shell of an atom:

Shell number	Number of allowed electrons
1	2
2	8
3	18
4	32
5	50
6	72
7	98

Fig 10-19 Number of electrons allowed in each electron shell of an atom

There are many ways that the periodic table can be drawn. Normally elements 57 to 70 and 89 to 102 are moved into a separate block. This has the advantage giving it a shape that takes up less space and means that the blocks for each element can be made large so that more information can be written in them. Here we are more interested in the shape of the table so in the tables in Fig 10-20 and Fig 10-21 those two sets of elements have been left in place.

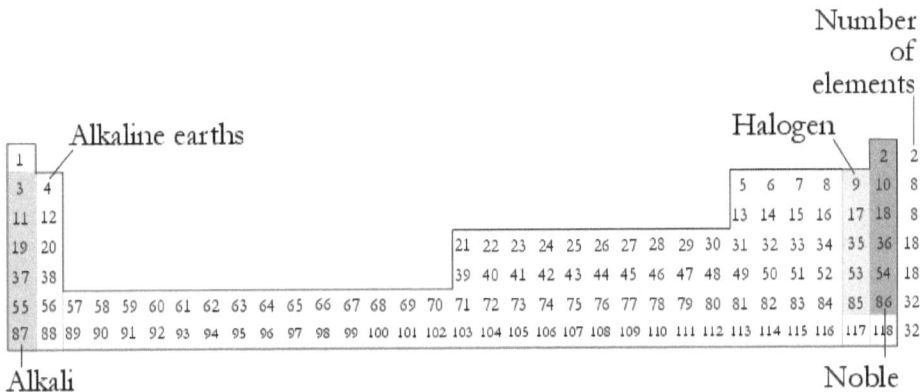

Fig 10-20 Periodic table showing atomic numbers

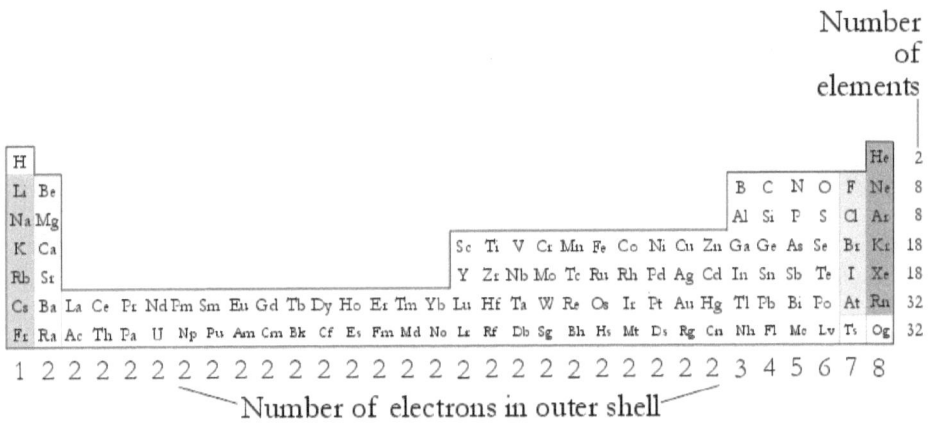

Fig 10-21 Periodic table showing chemical symbols

The shape of the periodic table reflects the number of electrons that each electron shell of an atom can accommodate. The top row contains just 2 elements: hydrogen and helium which have 1 and 2 electrons respectively. They can be held in the lowest energy shell of their atoms which can hold up to 2 electrons. The next row down has 8 elements and electrons in these atoms can be contained in the lowest 2 shells of their atoms.

Also, the first column holds elements with 1 electron in their outer shells and these elements all have the properties of alkali elements. The second column holds elements with 2 electrons in their outer shell and they have the properties of alkaline earths. The penultimate column holds elements whose outer shells contain 1 less than the maximum and they have the properties of halogens. And the right most column holds elements whose outer shells are full of electrons, the noble elements, whose atoms don't form bonds with other atoms.

The table looks like it has a jagged shape chunk missing because the electron shells nearest the nucleus cannot accommodate as many electrons as the shells further out which are used in the larger atoms. Another feature of the table is that it has two rows of 8 elements, two rows of 18 elements and two rows of 32 elements. Also, the number of electrons in the outer shell does not exceed 8. This does not seem to fit with the number of electrons allowed in each shell given in Fig 10-19. This is because the complex shape of orbitals causes some electrons from inner shells to have more energy than those in outer shells.

Elements 93 to 118 (Np to Og) are shown in a smaller font to indicate these are not found on earth they are only created in laboratories *Chapter 12 - Nucleus*.

Hydrogen is a bit of an oddity, it is not coloured brown because it does not always behave like an alkali element, but it fits best here in the table. Also, elements 117 and 118 have not been coloured because it has not been possible to test them. This is because only a few atoms of this type have ever been created and they have not existed long enough to be investigated.

So, the work and discoveries of Bohr, de Broglie, Pauli, Schrödinger and Heisenberg predicted and explained the shape of Mendeleev's periodic table along with all the properties of atoms which underlie chemistry.

These theories overcome the problems with Bohr's model of the atom. They predict the energy levels for atoms with any number of protons and electrons and in so doing they explain why each element has its own unique set of spectral lines (*Chapter 3 - Light part 1*). Also, the orbitals overcome the problem that the orbiting electrons should be radiating photons. The electrons do not orbit around the orbitals. They just exist somewhere within it. To be fair to Bohr he never used the term orbit to describe the atom. It was others who talked about electrons in this way.

The Heisenberg's theory is known as the uncertainty principle. It states that there is a lower limit to which we can be certain about the position of things. You might think "oh you just need better measuring equipment". But no. No matter how good your equipment is you can't determine the position of things better than this limit. It's as if the universe is a digital picture. When you zoom in you get to a point where you just see squares and you can't see what is in the squares[12]. The size of this limit is closely related

---

[12] To be more correct Heisenberg's uncertainty principle tells us that there is a limit to the precision with which we can know the position and momentum of a particle. The more precise

to the number which controls how light energy is chopped up into photons - Planck's constant (*Chapter 9 - Light part 2*).

This principle coupled with Schrödinger's equation determines the smallest size that electron orbitals can be. And this determines the size of atoms. This in turn determines the size of the smallest living cell. The combination of Schrödinger's equation and Heisenberg's uncertainty principle also explains why electrons do not fall into the nucleus.

Earlier success of Newton's laws of motion and the kinetic theory of gases had led people to think "if we know the current position, speed and direction of all particles we could work out what they do in the future and so we could foretell all of the future". This worked for planets and cannon balls but Heisenberg's uncertainty principle is telling us there is a limit to how precisely we can know a particle's position and speed so we can only give a range of possibilities for its future. If you put all the information about the universe into a computer with infinite power, it could only make limited predictions about the future.

The predictions of these wave equations sounded so weird that many people were reluctant to accept them, but these predictions turned out to agree precisely with the results of many experiments and they work for all atoms. There can be no doubt that electrons do have wave properties. Our modern technology is based on them.

So now we have a picture of an atom with a tiny but fantastically dense, positively charged nucleus which is surrounded by electrons which are confined to particular electron shells. For each shell there is a particular energy level. Each shell can contain one or more orbitals and each orbital can contain 1 or 2 electrons.

These arrangements of electrons explain chemistry and all properties of materials. The theories of Schrödinger, Heisenberg and Pauli explained and predicted how electrons behave this way in an atom. And it was realised that, without them, electrons would just be part of nuclei. And that would mean no molecules, so no compounds, so no rocky planets and no life.

Before the discoveries outlined in this chapter had been made, no link had been established between Chemistry and Physics. Afterwards they were closely united.

In later chapters we will see how these discoveries control which atoms can bond to form molecules and how energy is released from burning fuel. These theories also spawned the electronics which underpins modern technology.

---

we are about the position the less precise we can be about the momentum and vice versa.

Key points of this chapter

- Electrons in atoms are only allowed to have certain levels of energy known as energy levels
- When an electron falls from a higher energy level to a lower one it emits a photon whose energy matches the difference between the two energy levels
- Photons that hit an electron in a lower energy level will be absorbed and send the electron to a higher level
- The above will only happen if the energy of the photon matches the difference in energy between the two energy levels
- If an incoming photon has sufficient energy it will knock an electron out of the atom
- The above will cause the atom to become a positive ion
- The physical locations of these energy levels are known as electron shells
- Atoms of each element have a unique distribution of electrons among their shells
- Electrons fill the shell with the lowest energy level first, then, as more electrons are added they start filling the shell with the next higher energy level.
- Electrons in atoms have wave like properties that control the probability of finding them at a given location
- The shapes of these probability distributions are known as orbitals
- Each electron shell can contain one or more orbitals
- Each orbital can contain 0, 1 or 2 electrons, no more
- The above explains the shape and organisation of the periodic table
- The above explains and predicts the energy levels of atoms in all elements
- The positive charge and most of the mass of every atom is held in a nucleus at its centre
- The positive charge of an atom is tightly packed, and the negative charge is thinly spread out
- The positive charge in the nucleus is provided by particles called protons which have a mass nearly the same as the mass of a hydrogen

11 CHEMISTRY

Over centuries alchemists and scientists had been taking promising looking substances from the ground, mixing them with chemicals, boiling them and subjecting them to many different processes to produce useful substances and to extract elements. What they were doing was to split the bonds that held atoms together within molecules. Other scientists had been combining elements to form new compounds which often released heat or caused explosions. They were combining atoms to form molecules en masse.

At the end of the 19th century chemists were looking for answers:
- How do atoms bond?
- Why does the sequence halogen, noble, alkali, alkali earth repeat when elements are placed in order of atomic mass? (*Chapter 6 – Atom part 1*)
- Why do some pairs of elements combine to form compounds while others don't?
- Why do some reactions give off heat?

This chapter describes how knowledge about the structure of atoms covered in earlier chapters led to answers for these questions and how this has enabled chemists to produce compounds which have improved our understanding of biology and to make our lives more enjoyable.

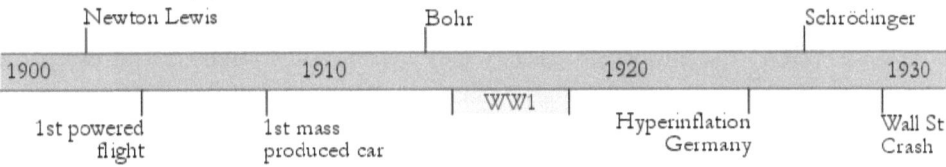

The great electron heist

In 1902 American chemist Gilbert Newton Lewis proposed that atoms bond together because one type of atom steals an electron from another type. For example, an atom of a halogen element will steal an electron from an alkali atom. If an atom steals an electron it will then have one electron more than it has protons and so it will have negative overall charge. It will become a negative ion. Similarly, the atom that lost an electron becomes a positive ion. So, because unlike charges attract, an electric force will pull the two atoms together - an effect known as ionic bonding. An example of this is sodium atoms bonding with chlorine atoms to form salt molecules.

But how can one atom steal an electron from another? For this to happen the attractive force between an electron and the donor atom's nucleus has to be less than the attractive force between that electron and the recipient atom's nucleus.

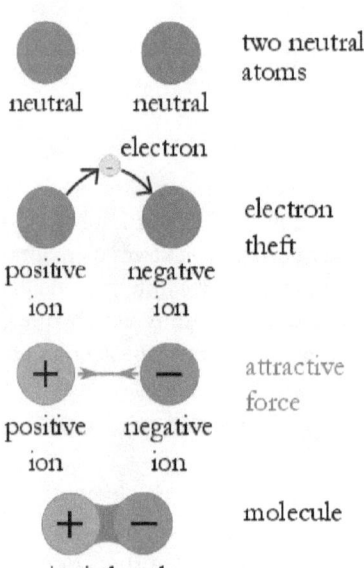

Two things affect the force between an electron and the nucleus: the distance between the electron and the nucleus and the way the electrons (and their charge) are distributed.

We saw in *Chapter 10 - Atom part 2* that electrons reside in a number of electron shells which surround the nucleus of each atom. An electron which is on its own in the outer shell of an atom will be relatively distant from the nucleus and so the nucleus will not attract it so strongly, so it could easily be lost from the atom, making it a positive ion.

What is harder to imagine is how an atom which has equal numbers of protons and electrons (making it electrically neutral) could attract and grab a nearby electron and become a negative ion. This is due to the way charge is distributed within an atom. The positive charge is concentrated tightly together in the nucleus while the negative charge is thinly spread out in the electron shells. In *Chapter 10 - Atom part 2* the analogy of the Albert hall and a gnat's abdomen was used to indicate just how spread out the electrons are compared to the protons in the nucleus. In *Chapter 2 - Charge in Solids* the tug of war analogy indicated how the same amount of charge packed tightly together exerts a stronger pull on a charged object than charge which is spread out.

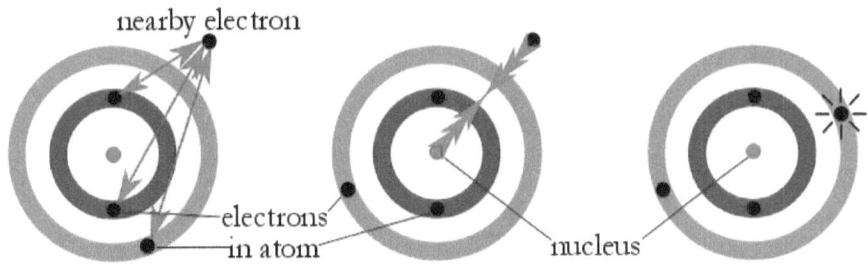

| repulsive forces between electrons in atom and nearby electron push in different directions | attractive forces between nucleus and nearby electron pull in the same direction | electron is drawn into a shell of the atom making it a negative ion |

Fig 11-1 How a neutral atom captures a nearby electron to become a negative ion

In this way, a free electron passing near an atom will feel a stronger pull towards the nucleus than the repulsive force offered by the other electrons in the atom. Therefore, it is very likely to get pulled into the atom, making the atom a negative ion. This can also happen to an electron which is in the outer shell of a nearby atom, turning one atom into a positive ion and the other into a negative ion. They bond together.

The strength with which atoms bond together governs how much heat is given off in chemical reactions and how strong a material is. It also affects the temperature at which elements freeze or boil as will be seen in *Chapter 13 - Atom part 3*. Atoms from different elements have different distributions of electrons among their electron shells (*Chapter 10 – Atom part 2*). In this way the distribution of electrons in the atoms controls the temperatures at which they boil and freeze.

In *Chapter 2 - Charge in Solids* it was mentioned that Volta had discovered that if two dissimilar metals come in contact with each other they each take on a different charge. One metal takes on negative charge and the other metal takes on positive charge. This is the phenomenon that his battery is based on. The way this happens is similar to the formation of ionic bonds (see above). The way electrons are distributed in the atoms of some metals mean they attract their outer electrons weakly and can easily lose them, whereas the distribution of electrons in the atoms of other metals means they will attract any nearby electron and that includes those in nearby atoms of other metals.

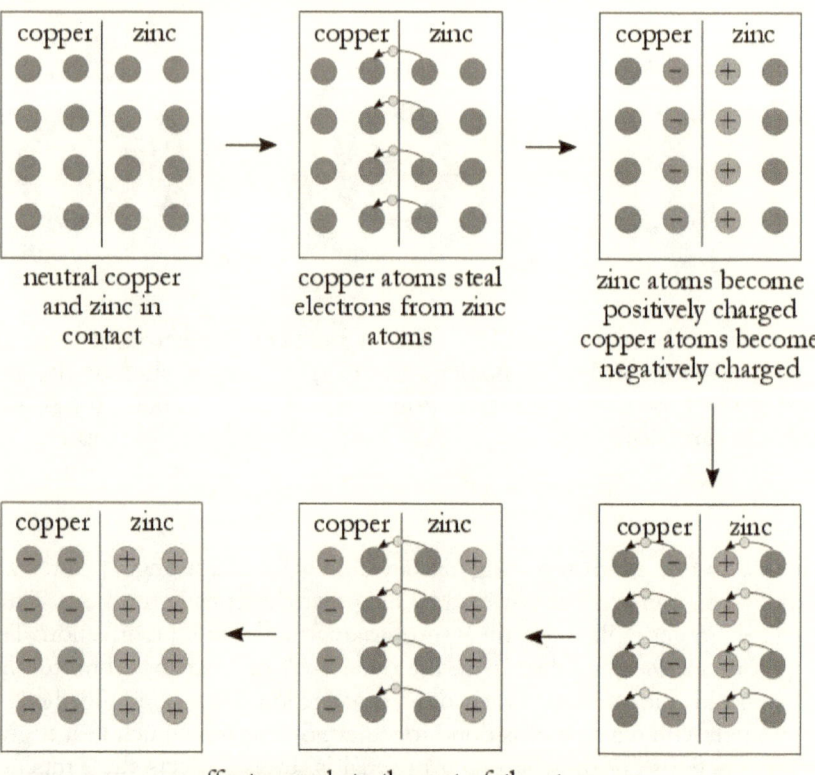

Fig 11-2 How copper and zinc gain their charges when they are in contact

In the case of the zinc and copper used in Volta's batteries the copper atoms steal electrons from the zinc atoms causing the first row of zinc atoms to become positively charged and the first row of copper atoms to become negatively charged. The next row of zinc atoms then lose an electron to the first row and the next row of copper atoms steal electrons from the first row of atoms. In this way the effect spreads into the two pieces of metal.

When Volta connected his wire across the other side of the pieces of copper and zinc excess electrons in the copper had a path to the zinc which was attracting them with its positive charge. And so the continuous flow of electrons ensues. This account is a little simplified to make it easier to describe the principle.

When atoms share electrons

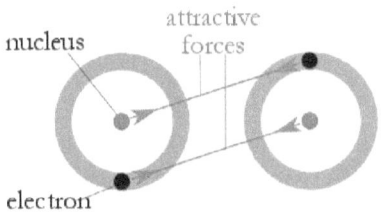

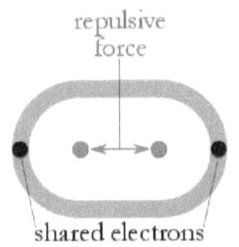

When certain types of atoms encounter each other there is a mutual attraction between them due to the attractive force between the nucleus of one and an electron in the other along with the attractive force between the nucleus of the second atom and an electron in the first. This causes them to approach each other and when they do the force of attraction between them increases. However, the repulsive force between the two nuclei also increases as they get closer. They approach until these two forces are equal. At this point the outer shell of the two atoms combine and share the two electrons. This is known as a covalent bond and it is this type of bond that holds two hydrogen atoms together in an H_2 molecule or hydrogen and oxygen together in a water molecule (H_2O). In some ways such a molecule is like an atom with more than one nucleus.

The arrangement of electrons within atoms

In *Chapter 10 - Atom part 2* we saw how Niels Bohr found that each atom is surrounded by electron shells and that each electron resides in one of those shells. He also showed that each shell had its own energy level and he calculated the energy of each energy level for a hydrogen atom which had only one proton and one electron.

As well as this we saw how Erwin Schrödinger produced his equation which predicts the energy levels of atoms with any number of protons and electrons. These electron shells comprise one or more orbitals and each orbital can accommodate one or two electrons but no more.

For atoms with multiple protons and electrons the number of electrons each shell can hold depends on the number of orbitals in that shell.

The work of Schrödinger, Pauli and Heisenberg showed that the first electron shell has one orbital, the second shell has four orbitals, the third has nine orbitals. These are shown in the left-hand diagram of Fig 11-3, the shapes in this diagram do not represent the actual shapes of the orbitals.

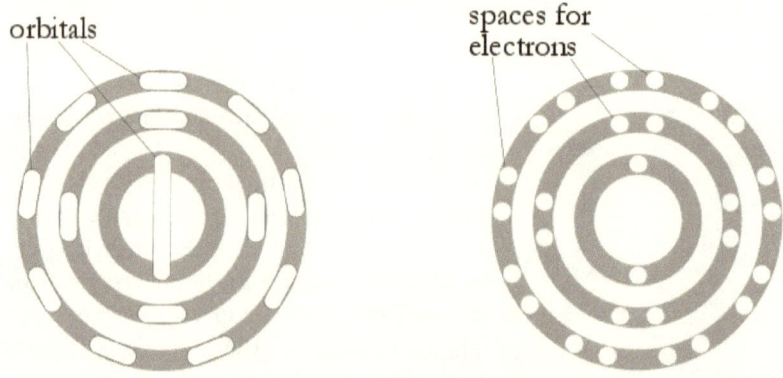

Fig 11-3 Orbitals and spaces for electrons in electron shells

Pauli had also established that each orbital can hold up to 2 electrons (*Chapter 10 - Atom part 2*), this means that the first shell can hold up to 2 electrons, the second up to 8 electrons, the third up to 18 electrons. The first three shells can hold 2, 8, 18 electrons as shown in the right-hand diagram in Fig 11-3.

The spooky thing is that the first four rows of the periodic table hold 2, 8, 8, 18 elements as mentioned in *Chapter 10 - Atom part 2*.

It was realised that electrons are attracted to the nucleus by the electric force and so they end up in the shell closest to the nucleus, which is the one where they have the least potential energy. But since there is a limit of two electrons in the first shell if the atom has 3 electrons (as is the case with lithium) the third one must go in the next higher shell. And so, the arrangement of electrons in shells for the lightest four elements will be as shown in Fig 11-3. Here full shells are shown in dark grey and part filled shells are shown in light grey.

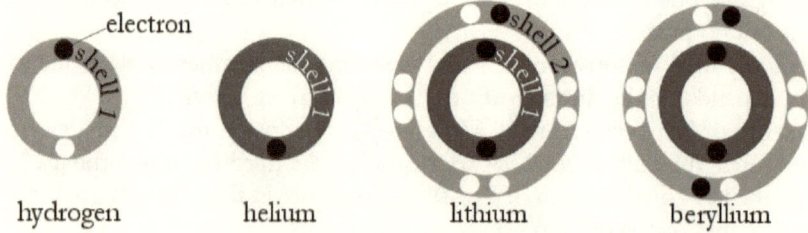

Fig 11-4 Arrangement of electrons in each shell for the lightest four elements

Element	Atomic number	Number of electrons in...			Group
		Shell 1	Shell 2	Shell 3	
Hydrogen	1	1	0	0	
Helium	2	2 (full)	0	0	noble
Lithium	3	2 (full)	1	0	alkali
Beryllium	4	2 (full)	2	0	alkaline earth
Fluorine	9	2 (full)	7	0	halogen
Neon	10	2 (full)	8 (full)	0	noble
Sodium	11	2 (full)	8 (full)	1	alkali
Magnesium	12	2 (full)	8 (full)	2	alkaline earth
Chlorine	17	2 (full)	8 (full)	7	halogen
Argon	18	2 (full)	8 (full)	8 (full)	noble

Fig 11-5 Table showing the number of electrons in each shell for selected elements

In Figures 11-4 and 11-5 the number of electrons in the partially populated outer shell of each atom is marked light grey and full shells are marked in dark grey. The same colours for the groups are used as in Figures 6-8 and 10-20. It becomes apparent that noble gases have a full outer shell[13], alkali elements have one electron in their outer shells, alkaline earths have two electrons in their outer shells and halogens are one electron short of a full outer shell.

The number of electrons in the outer shell determines the characteristics of each element and the shape of the periodic table. And these numbers of electrons are explained by the theories and equations of Schrödinger, Heisenberg and Pauli.

One thing that highlights the importance of outer electrons to the nature of elements is the comparison of lead and radon. It is natural to expect that radon, being a gas right

---

[13] Shell 3 can hold 18 electrons but it is deemed 'full' for Argon due to the complex shapes of orbitals.

down to -61°C would have lighter atoms but lead atoms have an atomic mass of 207 and radon atoms weigh in at 222. The thing is that radon atoms have full outer shells which makes it difficult for them to share electrons to form solids. More of this in *Chapter 13 - Atom part 3*.

Molecules also have an outer shell of electrons and these obey the same rules. Molecules that make up glass have an outer shell that only contains a few electrons and will easily lose an electron. Molecules of amber have an outer shell that is almost full of electrons so there is a tendency to steal electrons. So, when glass is rubbed it loses electrons and becomes positively charged and when amber is rubbed it steals electrons from whatever is rubbing against it and becomes negatively charged.

How rearranging atoms gives us heat

In *Chapter 1 - Energy* we saw that the temperature of a substance is a measure of how fast its molecules are moving or vibrating and the faster they move the more heat energy (kinetic energy) they have. Fundamental to our understanding of how burning creates heat is the idea of bond energy.

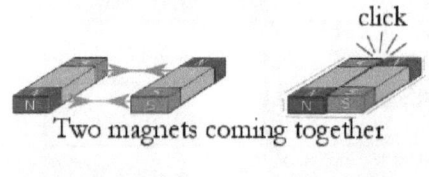
Two magnets coming together

To pull them apart considerable force must be applied

Bond energy can be pictured using two magnets. If you bring them close together you feel the attractive force between them. If you let go, they fly together and there is an audible click as they hit each other. That click is the equivalent of the bang when the rock hit the ground in *Chapter 1 - Energy*. Just like the rock hitting the ground and warming up, the magnets also get a little warmer. This click and warming is the bond energy of the two magnets.

Now, to pull them apart requires some effort. It is necessary to convert some chemical energy from your muscles to give the magnets sufficient kinetic energy to pull them apart against the magnetic force holding them together. This amount of energy is equal to the amount of energy that came out in the noise and warming as they clicked together.

This is known as the bond energy. When two objects that are attracting each other come together a bond forms and the bond energy comes out. To break the bond between two atoms it is necessary to apply them with an amount of energy equal to the bond energy. The same is true when two atoms that are attracting each other come together to form a molecule.

The strength of the attractive force between a pair of atoms determines the value of the bond energy. The stronger the force the more energy that must be applied to separate them and the higher the bond energy. Atoms from different elements have different distributions of electrons among their electron shells (*Chapter 10 – Atom part 2*) and this means the attractive force between different pairs of atoms will have different strengths and this in turn means the bond energies of different pairs of atoms will be different.

To extract heat from atoms all we need to do is to mix together some atoms which attract each other. Oxygen atoms are a good choice because they form strong bonds with atoms from other elements. The attractive force between hydrogen and oxygen atoms is very strong and the bond energy of these two atoms is high so they are a good choice if you want to generate some heat.

However, if we just mix oxygen and hydrogen together nothing will happen because, in the form we normally encounter them, these two gases comprise molecules of two atoms and there is no attractive force between these molecules.

Isolated hydrogen atoms are very rare on earth because as soon as they encounter one another they are attracted towards each other and they bond to form a molecule of two hydrogen atoms (H_2) which is the form of all of the hydrogen you are likely to encounter. The same is true for oxygen atoms. The oxygen which forms a quarter of the air you breathe is in the form of molecules of pairs of oxygen atoms (O_2). The only gases which naturally exist in the form of individual atoms are the noble gases and these are of no use in this respect because there is no attractive force between them.

To make hydrogen and oxygen atoms bond together it is first necessary to split the hydrogen and oxygen molecules into individual atoms. Adding a spark to the mixture of hydrogen and oxygen molecules will do this. Now the individual hydrogen and oxygen atoms can recombine to form molecules each with two hydrogen atoms and one oxygen atom: H_2O, water.

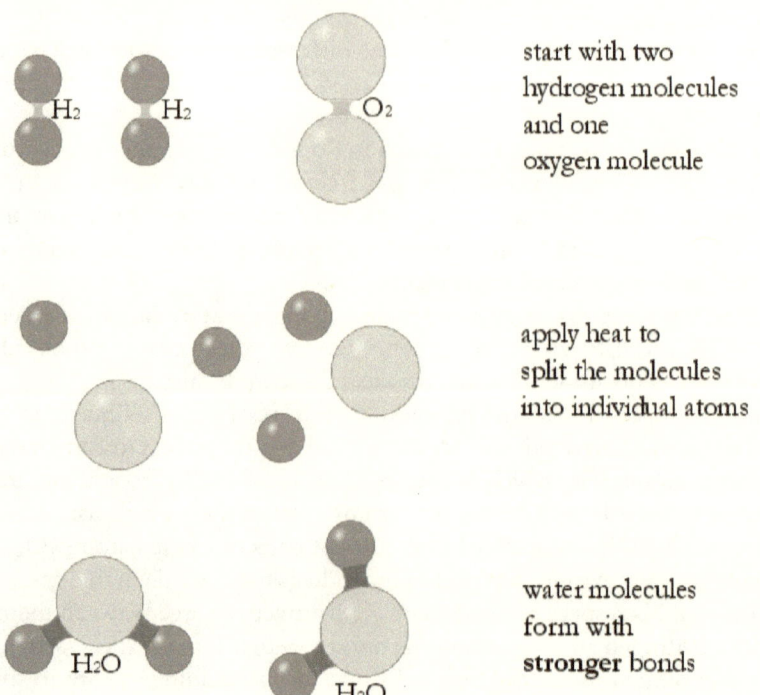

Fig 11-6 Weak bonds being broken, and strong bonds being formed as hydrogen burns to make water

The attractive force between hydrogen and oxygen atoms is much stronger than the attractive force between hydrogen atoms or between oxygen atoms. This means H-O bond energy is greater than the H-H and O-O bond energies. As the H-O bonds form to make H_2O, the energy released causes the H_2O molecules to fly about and collide with H_2 and O_2 molecules with more than enough energy to break these weaker bonds to create more free hydrogen and oxygen atoms. Consequently, a chain reaction occurs until all of the hydrogen or all of the oxygen has been consumed into water molecules. The sum of the bond energies of the H-O bonds created is less than the sum of the bond energies of the H-H and O-O bonds that were broken. Because of this, a large amount of heat energy is left over and so the gases get hot. As they get hot, they expand. In this case the expansion is so fast it would be termed an explosion.

We experience many situations where molecules split and recombine with oxygen to produce energy, for example, when we burn petrol in our cars or wood in our stoves. In muscles sugar and oxygen molecules in blood are split and recombined to produce different molecules and kinetic energy when you run or throw something. These are examples of chemical potential energy.

Oxygen is an element that, because of the way its outer electrons are arranged, bonds very strongly with many other elements. When anything burns, its atoms are bonding with oxygen atoms.

So, we have seen how the arrangement of electrons around an atom and the way concentrated charge has more effect than distributed charge allows some atoms to steal an electron from another and form a bond with it. Also, we have seen how the wave-like nature of electrons controls the distribution of electrons around atoms and how this causes the periodic nature of element properties as predicted by Schrödinger's equation. This also explains why only certain pairs of atoms will bond with each other and gives the elements their chemical properties. In *Atom part 3* we will build on this to explain why elements change from solid to liquid to gas as you heat them and why different elements undergo these changes at different temperatures, along with why some substances are transparent.

Key points of this chapter

- The way electrons are distributed among the shells of an atom is controlled by Schrödinger's equation
- Atoms of different elements have different distributions of electrons among their shells
- The attractive force between nucleus and electron is less for electrons in outer shells
- A free electron will be repelled by the electrons of an atom and attracted by the nucleus
- The attractive force of the nucleus is greater because it is densely concentrated
- An atom with a less-than-full outer shell can attract and capture a free electron
- An atom with a less-than-full outer shell can attract and capture an electron that is loosely attracted to its atom's nucleus
- This causes the thief atom to gain negative charge and the victim atom to gain positive charge thus there is attraction between the atoms, and they bond to form a molecule. This is known as ionic bonding
- When atoms bond they become hotter
- To break a bond between two atoms energy has to be applied to the molecule, usually in the form of heat
- Many chemical reactions involve putting heat in to break bonds in order for stronger bonds to form as the atoms rearrange themselves into different molecules
- If weak bonds are broken and stronger ones are made, excess heat comes out
- An example of this type of reaction is called 'burning'

12 NUCLEUS

Studies of the atom had told us that it contained a nucleus, but the atom still held mysteries. Why are atomic masses not round numbers and why are some elements radioactive? Investigations into the nucleus solved these mysteries and provided new sources of energy for mankind to use and to be fearful of, as we will see in this chapter.

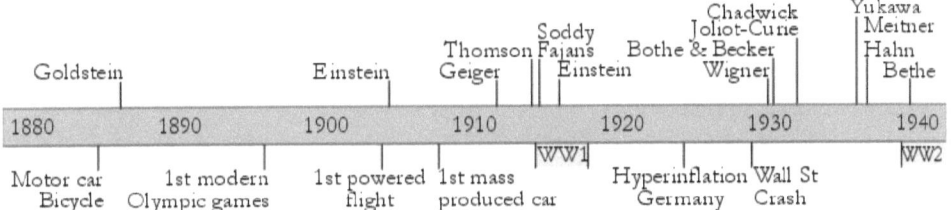

Rays at the other end of Crookes tube provide clues

In *Chapter 8 - Charge in Fluids* we saw that JJ Thomson made a small hole in the anode of his Crookes tube and investigated the small spot this produced on the surface of the tube.

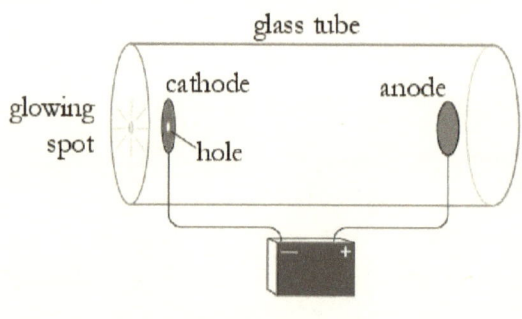

In 1886 German Scientist Eugen Goldstein did the opposite by using a Crookes tube with a small hole in its cathode (negative electrode). He found that a spot of fluorescence appeared on the glass near the cathode. In 1907 he decided to see what would happen when these rays were sent through a magnetic field just as Thomson had done. Unlike Thomson's spot caused by the electron, this one split out into several spots. From this Goldstein concluded they were from particles that were not all the same mass and named them canal rays. Also, unlike the spot caused by the electron, a different set of spots were observed when different gases were present in the tube.

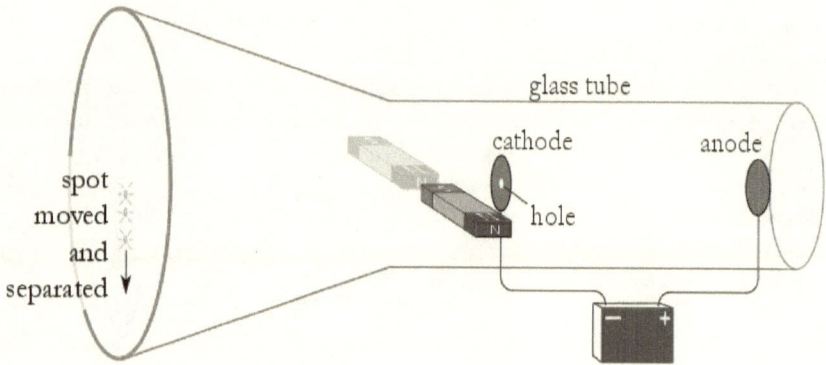

Fig 12-1 Goldstein's adaptation of a Crookes tube

From the way the spots moved in the presence of the magnetic field it became clear they were caused by the ions which Faraday had found in his experiments sending electricity through liquids. These were positive ions, atoms from the gas in the tube that had one or more electrons removed giving them a positive charge. Those electrons that had been removed from gas atoms were the ones that went flying off towards the anode.

So why should the magnetic field cause ions with different masses to take different paths and hit the screen in different places? It is a bit like threshed wheat being thrown up in the air on a breezy day. The light husks are affected more by the breeze and land further downwind than the heavier kernels.

This turned out to be a powerful technique for measuring and comparing the masses of particles and is widely used today going under the name of mass spectrometry.

In 1913 JJ Thomson developed this experiment further and arranged for these canal rays to fall on a photographic plate so he could study them more easily. He identified that two of the marks on the plate were from neon gas he had put in the tube. This was evidence that there were two types of neon atom with different atomic masses. He worked out that one of these spots was for neon atoms with a mass close to 20 protons and the other for those close to 22 protons. Regular Neon has 10 protons but an atomic mass of 20.18 protons.

At this time Scottish scientist Frederick Soddy, working at Glasgow university, was studying alpha and beta radiation of uranium and several other heavy elements. He found that the alpha rays were in fact the nuclei of helium which have two protons. He also found that as uranium gave off alpha rays it changed into thorium which itself gives off alpha rays and changes into radium. Atoms of these three elements have 92, 90 and 88 protons respectively. All this work involved a lot of careful investigation into the mass of the atoms involved and he concluded that atoms with the same number of protons can have different masses - the same conclusion as JJ Thomson but by a different means. The discoveries of Soddy were also made at the same time by Kazimierz Fajans in Poland.

So different nuclei with the same number of protons can have different masses. It seemed there must be something else in each nucleus which was causing the mass of atoms to vary.

A new subatomic particle solves old mysteries

In 1911, German scientist Hans Geiger found that if he filled a glass tube with a noble gas such as neon and put a source of alpha rays near it, he could see pulses of current on an ammeter monitoring the current going through the Crookes tube. By rearranging this with the negative electrode as a hollow tube and the positive electrode as a wire lying at the middle, he created a device that was far more convenient for detecting alpha rays than a scintillator (*Chapter 10 – Atom part 2*). This became known as a Geiger tube and with the addition of electronics to measure the rate of arrival of alpha rays the Geiger counter was invented.

When an alpha particle hits a neon atom in a Geiger tube an electron is knocked off the neon atom which becomes a positive ion. This positive ion is attracted to the negative electrode and thus on to the battery. So, this is a tiny blip of electrical current which is detected by the counter electronics.

The Geiger counter has become an essential tool in all nuclear physics work, but its first big contribution was in 1930 when German physicists Walther Bothe and Herbert Becker placed some beryllium near a piece of polonium that was emitting alpha rays.

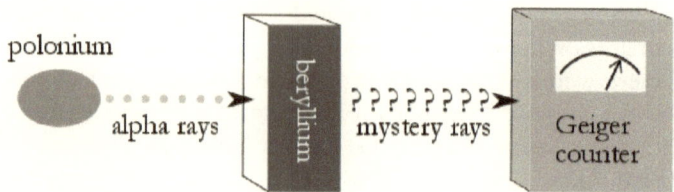

Fig 12-2 Discovery of mystery rays emanating from beryllium bombarded by alpha rays

They found that this caused the beryllium to emit some radiation which they detected with their Geiger counter. They tried the effect of putting different substances between the beryllium and the Geiger counter and found these rays would penetrate anything, including lead. So, these were not the alpha rays coming from the polonium because those could have been stopped by paper. They tried the effect of passing these mystery rays through magnetic and electric fields to see if these rays possessed charge: they didn't. Bothe and Becker were not certain, but they assumed they were gamma rays.

In 1932 Irène Joliot-Curie (daughter of Marie and Pierre Curie) and her husband Frédéric investigated what happened when this radiation hit the simplest of atoms, hydrogen atoms. The most convenient way for them to arrange hydrogen in their experiment was to use paraffin wax which has a lot of hydrogen in its molecules. They found the paraffin wax emitted protons when it was hit by the rays coming from the beryllium. By careful analysis they found the energy of the rays hitting the paraffin and emitting protons was far greater than had ever been seen in gamma rays suggesting they were something else.

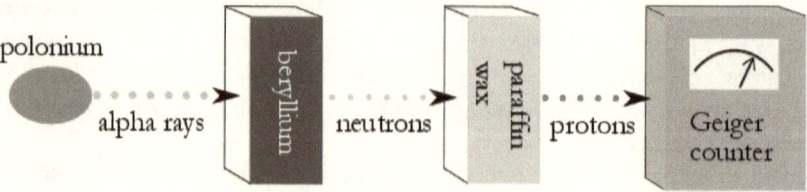

Fig 12-3 Identification of neutrons

Meanwhile British physicist James Chadwick developed an accurate Curie result he set up a similar experiment using polonium that had been given to him by Lise Meitner and applied this technique to the protons the paraffin emitted. He was able to show that the protons had been knocked out of the paraffin by a neutral particle with mass almost identical to the proton mass. He had discovered the neutral particle that Rutherford had proposed: the neutron.

Now the list of subatomic particles had increased to 3: electron, proton, neutron.

Neutrons and protons are the first two members of a group of particles known as hadrons to be discovered. Because they are the main constituents of the nucleus, neutrons and protons are also known as nucleons.

Well done James Chadwick, have a Nobel prize, you too Walther Bothe (sorry Herbert Becker, no prize for you).

The work by Goldstein, Thomson, Soddy and Fajans had shown that different atoms of the same element, while chemically identical, can have different masses and the nuclei have about twice the mass you would expect for the number of protons known to be there. The neutron explains all of this. The chemistry of an atom (what atoms it will bond with) depends on the number of protons because this controls the number of electrons in the outer shell. The number of neutrons makes no difference, so the number of neutrons can vary, changing the mass without changing the chemistry. There are about the same number of neutrons in a nucleus as protons so that is why the nucleus is about twice as heavy as the mass of the protons.

The neutron also explains the non-integer atomic masses of the elements as follows. A collection of atoms all with the same number of protons and neutrons is known as an isotope. Different isotopes of an element all have the same chemical properties. That is, they all do the same chemical reactions to form the same molecules because they have the same arrangement of electrons. When an oxygen atom bonds with a hydrogen atom it doesn't care how many neutrons it has, only the number of electrons in the outer shell is important. And that is determined by the number of protons. The number of neutrons in an atom has no effect on how it bonds with other atoms and so has no effect on chemistry.

If you obtain a lump or bottle full of some element (e.g. hydrogen) it will normally contain the same ratios of isotopes. Just like if you buy a bottle of semi skimmed milk it will always contain 1.8% fat. Your bottle of hydrogen will contain 99.98% atoms with 1 proton only, 0.0026% with 1 proton and 1 neutron and traces of hydrogen atoms with more than 1 neutron. The percentage of isotope that is naturally present in an element is known as its abundance. And this is why the atomic mass of hydrogen is 1.008 rather than 1.000[14].

Here is another way to look at it. A swimming pool of water made up of hydrogen and oxygen (H_2O) will have about a glass of water whose hydrogen atoms have 1

---

[14] An atomic mass of 1 is defined as one twelfth of the mass of a carbon atom. Carbon is used to define atomic mass because it is a solid at room temperature and so it is easier to determine its mass than hydrogen which is a gas at room temperature. This is another reason why the atomic mass of hydrogen is not 1.000.

neutron. The rest of the hydrogen atoms will have no neutrons. Water made from hydrogen with one neutron is known as heavy water. Heavy water can be used to slow neutrons down and provides a convenient way to control the speed of neutrons. This is of interest because many nuclear reactions only occur when the speed of neutrons hitting nuclei is just right. More of this later.

This varying number of neutrons is true for all elements. Each element comprises several isotopes and the atomic mass is the average of the isotopes taking their abundance into account.

Neon: atomic number: 10
        atomic mass: 20.18

Isotope			Number of Nucleons	Abundance
neon 20	10 protons	10 neutrons	20	90.48%
neon 21	10 protons	11 neutrons	21	0.27%
neon 22	10 protons	12 neutrons	22	9.25%

Fig 12-4 The three stable isotopes of neon give it its atomic mass of 20.18

All neon atoms have 10 protons, some have 10 neutrons (known as neon-20), some have 11 (neon-21) and some have 12 neutrons (neon-22). Normally there are more atoms with 10 neutrons than with 11 or 12. The abundances of these isotopes are neon-20 - 90.48%, neon-21 - 0.27%, neon-22 - 9.25%. This averages out to 20.18 - the same as the atomic mass for neon measured by chemists. The table below lists the stable isotopes of a number of elements. The term stable isotope means one that is not radioactive and does not change.

Isotope	Number of protons	Number of neutrons	Abundance
Carbon-12	6	6	98.9%
Carbon-13	6	7	1.07%
Nitrogen-13	7	6	99.6%
Nitrogen-14	7	7	0.4%
Oxygen-16	8	8	99.76%
Oxygen-17	8	9	0.04%
Oxygen-18	8	10	0.2%
Fluorine-18	9	9	trace
Fluorine-19	9	10	100%
Neon-20	10	10	90.48%
Neon-21	10	11	0.27%
Neon-22	10	12	9.25%

Fig 12-5 Stable isotopes of some common elements

The elements with atomic numbers 1 to 82 (hydrogen to lead[15]) have a number of stable isotopes and trace amounts of some unstable isotopes. Elements heavier[16] than lead have no stable isotopes. Unstable isotopes are radioactive and are constantly changing into other elements by giving off alpha or beta radiation. For example, the isotope neon-23 (10 protons and 13 neutrons) gives off beta- radiation. When this happens, a neutron turns into a proton and the atom becomes sodium-23 (11 protons and 12 neutrons). After 37 seconds half of any sample of neon-23 will turn into sodium-23. Another 37 seconds later, half of what's left will turn into sodium-23. These 37 seconds are known as the half-life. It could be said that a mug of tea has a half-life of 15 mins. That's how long it takes to cool to half-way between its starting temperature and room temperature. Over the next 15 mins it will cool by half of that. Eventually it will reach room temperature.

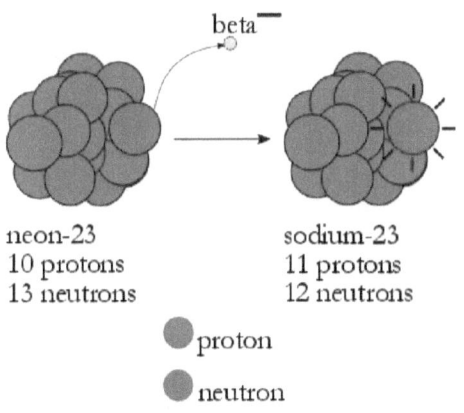

neon-23
10 protons
13 neutrons

sodium-23
11 protons
12 neutrons

● proton
● neutron

---

[15] There are two exceptions – [43]Technetium and [61]Promethium which have no stable isotopes
[16] A heavier element is one which has more protons in its nuclei

The decay of neon-23 into sodium-23 is an example of beta- radiation. The other type of beta decay happens when carbon-11 decays into boron-11 and emits a beta+ particle. The beta+ particle is also known as a positron which is used in hospitals PET scans (Positron Emission Tomography).

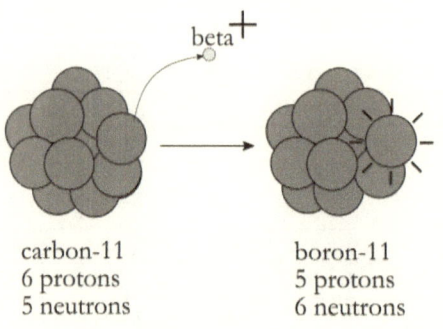

carbon-11
6 protons
5 neutrons

boron-11
5 protons
6 neutrons

The effect these different types of radioactive decay have on nuclei is shown in Fig 12-6. The emission of a gamma ray does not change the number of protons or neutrons in a nucleus, it accompanies alpha and beta radiation to dispose of excess energy from the nucleus.

Type of radioactive particle emitted	Effect on the nucleus	Effect on isotope
Alpha	Lose two protons and two neutrons	Changes to element with atomic number - 2
Beta⁻	Neutron turns into a proton	Changes to element with atomic number + 1
Beta⁺	Proton turns into a neutron	Changes to element with atomic number - 1
Gamma	Loses energy	No change

Fig 12-6 What happens to nuclei when they emit alpha or beta particles

For each element there are a few numbers of neutrons that make the nuclei stable. For example, carbon isotopes are stable if they have 6 or 7 neutrons, the same for nitrogen, for oxygen it is 8,9 or 10 neutrons. Carbon isotopes with less than 6 neutrons or more than 7 neutrons are unstable. Oxygen isotopes with less than 8 neutrons or more than 10 neutrons are unstable. For each element there is a minimum and a maximum number of neutrons that make the nucleus stable. Nuclei with numbers of neutrons less than the minimum or greater than the maximum are unstable and will emit some particle of radioactivity (alpha, beta+ or beta⁻). This means they turn into a different isotope as outlined in Fig 12-6[17].

So, does this mean everything is radioactive? Yes, but don't worry, the amounts of radiation from most things we come across in normal life is too small to worry about. The most radioactive thing you are likely to come across in day-to-day life is another person.

Uranium with atomic number 92 is the heaviest element found on earth. Elements heavier than uranium are known as transuranic elements. In *Chapter 14 - Cosmology part 2* we will see why we don't find these elements on earth.

A dating agency is opened

As can be seen in Fig 12-5 a regular piece of carbon comprises 98.93% carbon-12 (6 neutrons), 1.07% carbon-13 (7neutrons). It also has a trace amount of carbon-14 (8 neutrons) which is unstable and emits negatively charged beta⁻ radiation in the same way that neon-23 did. As it does so it turns into an isotope of nitrogen with 7 protons and 7 neutrons.

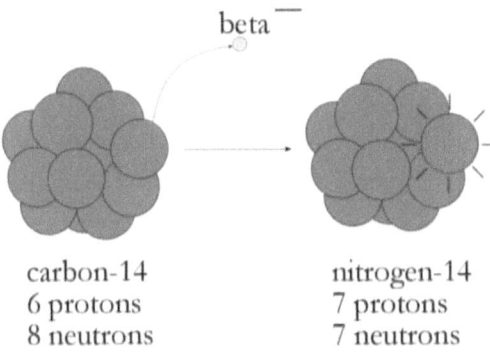

carbon-14
6 protons
8 neutrons

nitrogen-14
7 protons
7 neutrons

Carbon-14 has a half-life of 5715 years. It reduces by half every 5715 years. What is happening is that every so often a neutron turns into a proton and when it does so it emits a positive beta-particle. So, the carbon atom with 6 protons and 8 neutrons turns into a nitrogen atom with 7 protons and 7 neutrons.

Our atmosphere is constantly being bombarded by the nuclei which are thrown out by cataclysmic events throughout the universe known as supernovae (*Chapter 14 - Cosmology part 2*). Mostly these are hydrogen nuclei which are just single protons. These nuclei which are raining down on us are known as Cosmic rays.

It so happens that carbon-14 is constantly being created in our atmosphere by cosmic rays hitting nitrogen atoms and forming a tiny amount of carbon-14 which ends up in carbon dioxide (a molecule with 1 carbon atom and 2 oxygen atoms) in the atmosphere. Plants absorb this carbon dioxide while they are alive but stop doing so when they die.

---

[17] More recently, other forms of radioactivity have been discovered but they are less common.

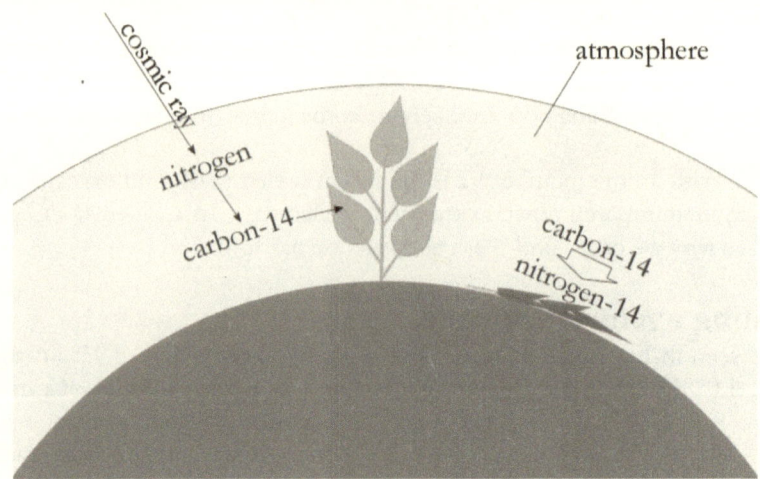

Fig 12-7 Plant absorbing carbon-14 while it is alive which slowly turns into nitrogen -14 after it dies

That means while they are alive, they are constantly being topped up with carbon 14. It doesn't affect them, but when they die the amount of carbon-14 they have starts to reduce at a predictable rate. Every 5715 years half of the carbon-14 (unstable) changes to nitrogen-14 (stable).

Since the 1940s it has been possible to measure the amount of carbon-14 in living things and dead organic matter accurately enough to detect how much the carbon-14 has decayed since that plant was living. By studying plants that are currently alive it is possible to determine the amount of carbon-14 they have when they are alive. So, if they find a similar thing that died a long time ago, knowing the rate that the carbon 14 turns into nitrogen and knowing how much carbon-14 it has now they can find out how long ago it died. It is possible to date things back 50,000 years in this way.

This carbon-14 we breathe is part of what makes humans the most radioactive thing you will probably encounter in normal life.

A similar process using isotopes of potassium and argon allows the ages of rocks to be determined. When planets are newly formed, they mostly comprise minerals that are so hot they are molten. As the molten minerals cool, they form crystals that coalesce as rocks. By the age of a rock we mean how long ago the atoms and molecules that make it up cooled sufficiently to form crystals[18].

From the ratios of isotopes in the various elements in some meteorites found on earth it has been possible to determine that they are older than our planet.

---

[18] There are 3 main types of rock: igneous, metamorphic and sedimentary. This is true for igneous rocks, a type of rock forms when magma, ejected by a volcano, cools and solidifies.

Splitting nuclei releases a lot of energy

So now the picture of the nucleus is that it contains positively charged protons and neutrons which have no charge. The number of protons will be matched by the number of electrons in the electron shells surrounding the nucleus. The arrangement of these electrons determines the chemical properties of the element (*Chapter 11 - Chemistry*). Therefore, the number of protons determines the chemical properties of the element.

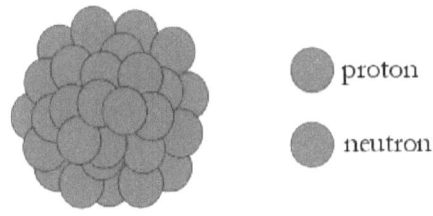

proton

neutron

Mostly there are a few more neutrons than protons in a nucleus. Neutrons are very slightly heavier than protons.

All very nice but this holds a mystery. The protons in every nucleus are all positively charged and so are repelling each other. The closer they are, the greater the repulsion. In the nucleus they are about as close as it is possible to get. So why do they not all fly apart? Hungarian-American physicist Eugene Wigner suggested that another force must be responsible for holding the nucleus together. This became known as the strong nuclear force. It holds the protons and neutrons tightly together in the nucleus even though protons hate being close together because they are all positively charged and so they are trying to repel each other.

In 1935, Japanese scientist Hideki Yukawa produced a comprehensive theory which describes how and why the nuclear force exists and why it acts over such a small distance.

So now we have a third force, nuclear. The first two, gravity and electromagnetic, are long-range forces whose effect reduces gradually the further things are apart. The nuclear force is not like that. It only affects protons and neutrons and has no effect on protons or neutrons that are further apart than the diameter of a nucleus. It is a short-range force.

In *Chapter 1 - Energy* we saw that when there are two forces acting on an object in opposite directions it will have potential energy. So that should mean each nucleus with multiple protons has potential energy.

This was confirmed in the 1930s when Austrian scientist Lise Meitner and German scientist Otto Hahn worked together in Berlin on experiments exposing uranium to streams of neutrons. Mainly Lise Meitner carried out the theoretical work and Otto Hahn did the chemical analysis to detect what elements had been produced in their experiments. In 1938 Lise was forced to move to Stockholm because of her race but she continued to collaborate with Otto. The intention of these experiments was to show that the extra neutrons knock bits off the uranium (92 protons) nuclei and create another element a few protons lighter such as radium which has 88 protons. After Lise had left, Otto continued doing experiments suggested by Lise in letters she wrote to him. Otto found barium (56 protons) and krypton (36 protons) were being produced. This was not at all what was expected because no one believed the nucleus could be split almost in half.

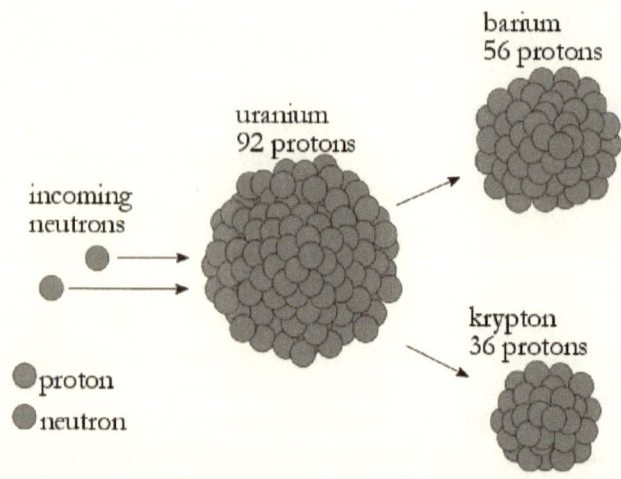

Fig 12-8 Splitting of uranium nuclei into barium and krypton nuclei

The likelihood of these reactions happening depends on the speed of the neutrons hitting the uranium. Too slow or too fast and the neutrons are not absorbed, and the uranium does not split. Meitner and Hahn used heavy water to slow down the neutrons coming from the radioactive source.

One way to think of this is to imagine a well surrounded by a mound. A ball is rolled towards the well. If it is going too slowly it will go a little way up the slope of the mound and roll away again. If it is too fast, it will go up the slope of the mound and keep going and overshoot the well. If it has just the right speed it will go over the mound and plop into the well.

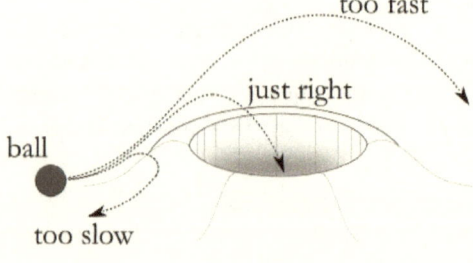

Lise extended Yukawa's work and produced a theory to show how the addition of a neutron to a uranium nucleus could disrupt the strong nuclear force that was holding the protons in the nucleus together and showed that the uranium nucleus had split almost down the middle. When this happens the electrical repulsive force between the two new smaller nuclei would cause them to fly apart with great speed. These objects now had a very large kinetic energy. This action is called nuclear fission.

Protons in a nucleus feel two forces: the electric force pushing them apart and the nuclear force holding them together. In a stable nucleus the nuclear force wins and holds the protons very tightly together. In the case of Hahn and Meitner's uranium nucleus hit by a neutron the nuclear force was disrupted and became unable to hold the protons together. The electric repulsive force sent two chunks of the nucleus in the form of barium and krypton nuclei flying off at great speed. Crucially, as we will see later, some lone neutrons went flying off too.

The reason the two halves of the nucleus departed so quickly was that the closer they are together the greater the repulsive force between them caused by the fact that they have positive charge. Inside a nucleus they are held incredibly close together by the nuclear force. You could think of the nucleus as a balloon holding balls together which are trying to fly apart. When the balloon breaks, two sets of balls fly off in opposite directions.

In *Chapter 14 - Cosmology part 2* we will see how this nuclear energy got into the nucleus.

At first it was thought the energy produced by splitting uranium into two nuclei about half the size would be too difficult to harness. However, it was also found that when these nuclei split it was not a clean split. A few neutrons went off on their own. If these could be made to cause other uranium nuclei to split, a chain reaction would ensue, and all the uranium atoms would be split into smaller atoms such as barium and krypton and a large number of atoms would be flying apart very quickly. You would have an atom bomb.

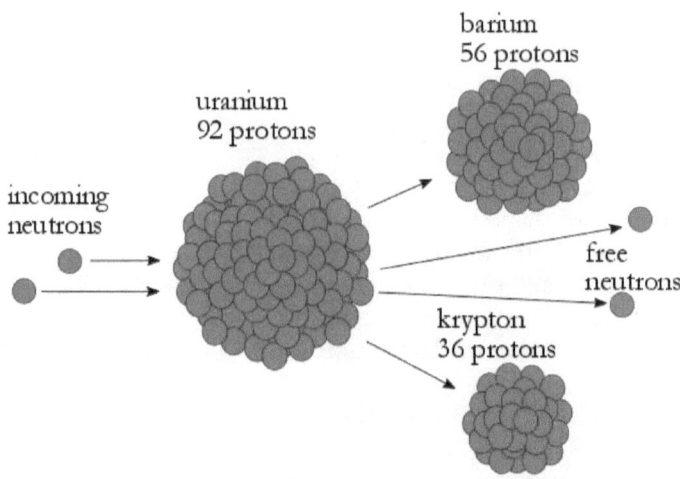

Fig 12-9 Splitting of uranium nuclei into barium and krypton nuclei with release of free neutrons

In a chemical explosion there is a chain reaction caused by the heat released by one molecule making adjacent ones hot enough to burn as in *Chapter 11 - Chemistry*. In nuclear fission the chain reaction is caused by the neutrons that are released when one nucleus splits then hitting an adjacent nucleus.

Meitner also calculated the masses of the nuclei and found that the sum of the masses of the barium, krypton and free neutrons was less than the mass of uranium they started with by about one fifth of a proton. She plugged this missing mass into the m part of Einstein's famous equation $E=mc^2$. The result was the same as the kinetic energy of the barium, krypton and free neutrons.

Anything that has potential or kinetic energy has some relativistic mass. Under normal circumstances relativistic mass is far too small to measure. In a nuclear fission reaction however, as Meitner had discovered, the amount of energy is so large that the relativistic mass becomes a significant proportion of the regular mass. Einstein's relativity tells us that if you managed to get an object up to the speed of light its mass would become infinite, but we don't need to worry about that. That would require an infinite amount of energy.

Many accounts of nuclear fission say that a bit of mass has been converted into energy which sounds nice and spooky. But it is confusing and not quite right. No particles were annihilated or harmed in any way. When the barium and krypton were weighed, they had cooled down. They had lost their kinetic energy and so they had lost their relativistic mass. That is why they weighed less. The law of conservation of energy had been obeyed.

So, relativity is not required to understand nuclear fission, but as Lise Meitner showed, nuclear fission does confirm relativity in that it shows energy has the equivalent mass predicted by Einstein.

Or, to put it another way, no mass was turned into energy, but it did show that energy has equivalent mass.

Later experiments showed that all elements heavier (more protons per nucleus) than iron can be made to fission and will release energy in doing so. Elements lighter than iron can also be made to fission but you have to put more energy in than you get out.

We have encountered the nuclear force which holds protons together in the nucleus. But it was realised that another, weaker force must exist in the nucleus - a force which causes alpha beta and gamma rays to come flying out of the nucleus. This has been named the weak nuclear force. The force which holds protons together has been named the strong nuclear force.

Bonding nuclei releases even more energy

The opposite of nuclear fission is nuclear fusion where two nuclei are brought together, and they bond to form a new larger nucleus. However, this is not easy to do. As the two nuclei come closer together their positive charge repels them. This repulsion gets greater the closer they are together. However, if they do get close enough together the strong nuclear force takes over and they bond and form a new larger nucleus. The way to do this is to throw them together very fast so their kinetic energy is enough to overcome the repulsive force of their positive charges. We saw in *Chapter 1 - Energy* that temperature is a measure of how fast a body's particles are moving. If you make hydrogen gas very hot the molecules will fall apart and then the electrons will be stripped off the atoms. At a temperature of around 100 million °C you will have some hydrogen nuclei moving at great speed.

When two nuclei do bind together, the nuclear force that pulls them together is stronger than any other known force. They end up moving together very fast. Fast moving particles is what heat is. A lot of heat energy comes out when this happens. This is just like the rock falling off the cliff. When it hit the car, its kinetic energy was converted to heat and other vibrations.

In 1939, German scientist Hans Bethe worked out that hydrogen in the centre of stars is hot enough for this to happen. Hydrogen nuclei with one proton are fused together to make helium (2 protons). Then enough heat is generated to make other hydrogen nuclei do the same. This was the first successful explanation of how stars shine. Others showed how 3 helium atoms are fused to make carbon (6 protons) in stars. Eventually it was found that all atoms are formed that way in stars as will be seen in *Chapter 14 - Cosmology part 2*. That includes those that make us and everything around us. This is not a tidy process; all manner of unstable isotopes and spare neutrons are created. The unstable isotopes produce all forms of radioactivity which along with spare neutrons leave the star to form the cosmic radiation that permeates space and our atmosphere.

Later calculations showed that, as with fission, all elements lighter (less protons) than iron can be made to fuse and will release energy in doing so. To fuse elements heavier than iron you have to put more energy in than you get out. Bethe had discovered not only how stars shine but also suggested a process for how and where elements are made.

So light atoms can be fused together to make heavier atoms and in doing so they release a lot of energy. Unfortunately, this allows us to make H bombs. On the plus side, experiments are ongoing worldwide which will, one day, allow the conversion of vast amounts of nuclear energy into usable electrical energy when isotopes of hydrogen are joined to form helium atoms in the nuclear fusion process. For example, the Joint European Torus (JET); International Thermonuclear Experimental Reactor (ITER); National Ignition Facility (NIF) etc. The source of the hydrogen will be cheap and plentiful water.

Studies of the nucleus have shown us why the atomic masses of elements are not round numbers, the basis of radioactivity and how to release enormous amounts of energy from the nuclei of atoms. Nuclear fission has been harnessed to convert potential energy in the nucleus to electrical energy in nuclear power stations but it brings with it so many problems people ask if it is worth it. In future, nuclear fusion will allow far greater amounts of nuclear energy to be converted to electrical energy. The challenges are greater but so are the potential rewards. Also, the problems will be far less. Electrons, neutrons and protons make up just about everything we see from the slime that allows snails to move to the diamonds in a monarch's crown. Slime and diamonds are made of the same protons, neutrons and electrons. They are different only because their protons, neutrons and electrons are arranged in different ways.

Key points of this chapter

- A nucleus comprises two types of particle: protons and neutrons and these provide most of the mass in an atom
- The proton has positive charge and the neutron has no charge
- The neutron has slightly more mass than the proton
- All atoms in an element have the same number of protons
- Atoms in an element have varying numbers of neutrons
- The varying number of neutrons in atoms is the reason why the atomic mass of elements are not integer numbers
- A collection of atoms which all have the same number of protons and the same number of neutrons is called an isotope
- Some isotopes are radioactive which means they emit radiation in the form of alpha, beta, gamma or neutron particles
- When a nucleus emits a beta particle a neutron changes into a proton or a proton changes into a neutron causing them to become a different element
- The rate at which neutrons turn into protons allows people to find the age of organic matter by examining the amounts of certain isotopes present
- The protons are packed extremely tightly together in the nucleus and this gives them enormous amounts of potential energy because they are repelling each other
- A very strong nuclear force holds the protons and nucleons extremely close together in a nucleus
- An incoming neutron can cause a nucleus to split apart and cause a large amount of this potential energy to change into kinetic energy as the leftover parts of the nucleus go flying off
- If two small nuclei are forced close together, they gain an enormous amount of potential energy as the strong nuclear force starts to attract the other nucleus
- This potential energy converts to kinetic energy as the two nuclei 'fall' together and this causes an enormous amount of heat energy to be released

13 ATOM PART 3

Science has delivered an intricate model of how and why atoms and light interact in the way they do. This chapter shows how this model explains why we have solids, liquids and gases, what they are like inside, why some things are transparent, why some things conduct electricity and other such things.

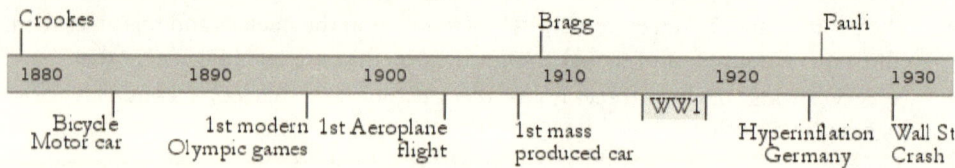

Crystals

Crystals can form when certain liquids evaporate such as salt water or sugar water. Other types of crystal can form when hot liquid metals or rocks solidify as they cool.

Crystals can vary in size from those that can only be seen in a powerful microscope to giant 4 metre gypsum crystals found in the Niaca mine in Mexico. The size of crystals depends on the rate at which they form, in general the slower they form the larger the resultant crystal. Some iron crystals in meteorites are so large, calculations show they must have cooled at a rate of thousands of years per °C which would not have been possible on earth.

The formation of crystals is also affected by the pressure at which they form. Diamond only forms when carbon freezes under intense pressure deep in the earth. Under other circumstances you end up with graphite. There are several different forms of water ice. The type which forms snowflakes depends on the pressure and the humidity of the air when water vapour freezes.

Crystals pose lots of questions, to answer them scientists would need to know how the atoms are arranged in a crystal.

X-rays find the structure of crystals

Investigating crystals with light gave no indication of their structure, but a great step forward was made in 1912 when German physicist Max von Laue sent a beam of X-rays through a copper sulphate crystal. He found the X-rays passed through the crystal to produce a pattern of dots on a photographic plate placed behind it. He realised this was caused by diffraction just as Young and Fraunhofer had found when they shone light through slits and gratings (*Chapter 3 – Light part 1*). This time, rather than something man-made, the diffraction was caused by the regular spacing of atoms in the crystal. Von Laue knew the wavelength of the X-rays in his beam and from this he calculated the distance between atoms in the crystal which is about 0.3 nanometres. (A nanometre is 1/1,000,000,000th of a metre). Visible light does not diffract when it encounters a crystal because the shortest wavelength (450 nanometres) is much longer than the distance between the atoms.

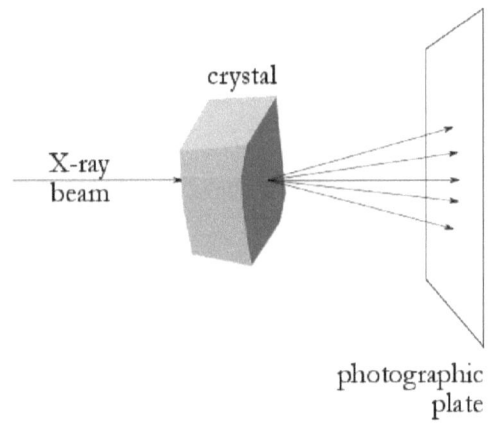

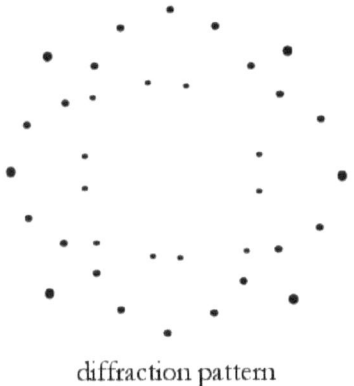

diffraction pattern of salt crystals

Further progress was made in 1914 when British scientists William Bragg and his son (also William), working at Leeds University, UK, reproduced and refined von Laue's work. They found the dots produced on the photographic plate were arranged in patterns of curves and they developed a mathematical technique which revealed that the atoms in salt were arranged in a cubic lattice structure as shown in Fig 13-1.

The left-hand diagram shows the chlorine (Cl) and sodium atoms (Na) which make up salt with their sizes in proportion to the size of the crystal. However, the atoms at the front hide the other atoms which makes it difficult to see the structure of the crystal. For this reason, diagrams of crystals are often drawn with out-of-proportion atoms as in the middle diagram. It can be seen that the lattice structure is made up from many cubes stacked on top and beside each other. These cubes determine the shape and structure of the crystal and are named unit cells as in the right-hand diagram.

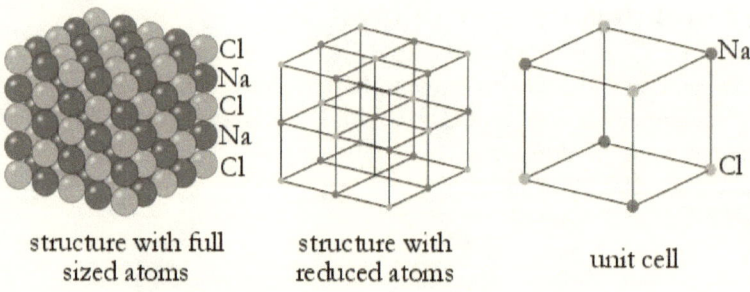

Fig 13-1 Representations of atoms in salt crystals

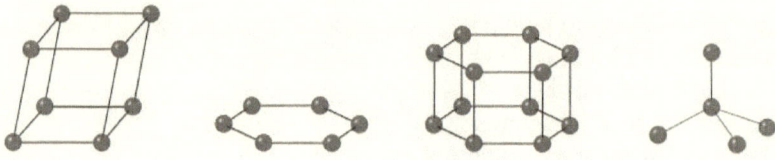

Fig 13-2 Examples of other types of unit cell

Fig 13-2 shows other types of unit cell which produce crystals with different shapes and hardness. The shape of unit cells mainly depends on the types of atoms involved. This is because it is affected by which orbitals in the outer shells of the atoms are populated with electrons and the strength of attraction between the atoms. This will be explored further in this chapter where we look at the shape of molecules.

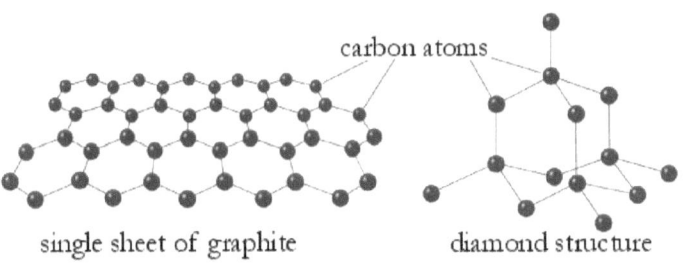

Fig 13-3 Two forms of carbon crystal

When carbon forms crystals it normally forms sheets of carbon atoms comprising hexagonal unit cells. Graphite comprises many of these sheets stacked on top of each other. But if carbon crystalises under great pressure, for example deep in the earth's crust, it forms into a diamond which has unit cells in the form of tetrahedrons.

The use of X-rays to find the structure of crystals is known as X-ray crystallography which has allowed scientists to probe the structure of nearly all crystals. In the 1950s chemist Rosalind Franklin used the Braggs' technique to discover the double helix structure of DNA.

Well done Max von Laue and William Bragg, have Nobel prizes.

Sorry Rosalind.

Why some substances are transparent

Up until now the electron shells and orbitals where electrons reside in atoms have been shown as sharply defined objects with hard edges. But Schrödinger's wave equations tell us that orbitals describe the possibility of finding electrons in a certain place around an atom. Further, they tell us that you are more likely to find the electron an orbital holds at the centre than further out. The likelihood of encountering an electron diminishes as you go out from the centre of the orbital. So, if orbitals are considered in more detail they appear more like fuzzy objects.

In *Chapter 11 – Chemistry* we encountered the covalent bond where atoms bond to form molecules by sharing electrons. Atoms in crystals bond in a similar way, only in this case instead of sharing electrons between a few atoms in a molecule the outer electrons of atoms are shared between millions of atoms in a crystal. So, they become a sort of fuzzy super orbital or electron cloud which permeates the entire crystal.

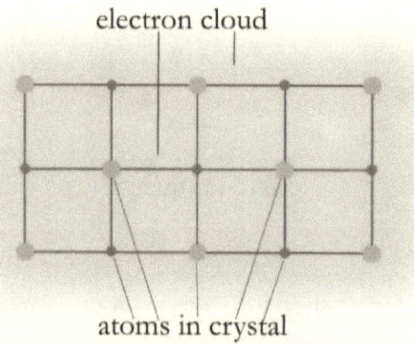

In *Chapter 10 – Atom part 2* it was established that Pauli's exclusion principle holds that no two electrons in an orbital can have the same set of quantum numbers. For the purposes of this explanation we can say that it is telling us that only two electrons in an orbital can have the same energy level. But in the crystal we have millions of electrons in the super orbital, so something is going to have to give way. What happens is that the energy levels of the electrons in the electron cloud all adjust their energy slightly so only two occupy exactly the same energy level. And this causes the energy level of the outer electrons to smear out into an energy band.

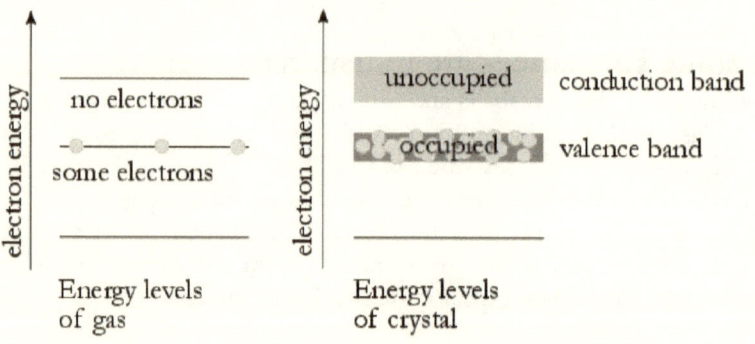

Fig 13-4 Energy levels and energy bands

As we saw in *Chapter 10 – Atom part 2* electrons fill the lowest energy level first then when that is full they start filling the next higher energy level. Here the highest energy level has smeared out into an energy band. The energy band with the highest energy that contains electrons is known as the valence band. The one above it which has no electrons is known as the conduction band. An electron of any energy in the valence band can be sent to any energy in the conduction band if it is hit by a photon of sufficient energy. This photon will be absorbed by the electron.

So, photons with a range of energies can be absorbed to push electrons between bands. Because the wavelength of a photon is proportional to its energy (*Chapter 9 - Light part 2*) this means the electrons in the crystal are much less fussy about the wavelengths of the photons they absorb. This range of wavelengths of photons is known as the absorption band. If this absorption band corresponds to the visible part of the electromagnetic spectrum the crystal will be opaque.

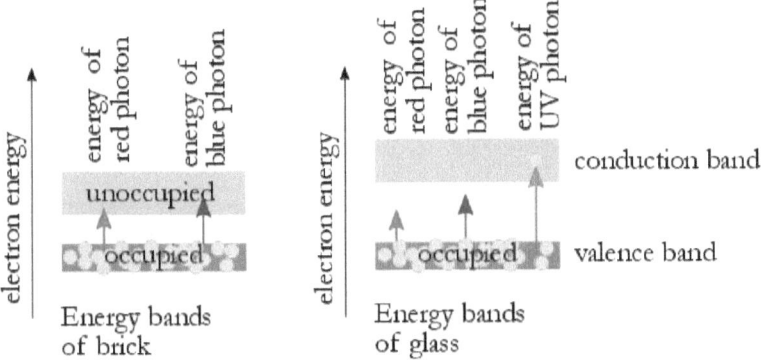

Fig 13-5 Gaps between energy bands determine whether objects are opaque or transparent

So that's why most solids are opaque. All visible light photons are absorbed by the electrons of the outer shells as they send electrons from the valence band to the conduction band. Glass is transparent because the energy gap between the top two energy bands is greater than the energy of photons of the visible part of the electromagnetic spectrum. This means that when a photon of visible light hits an electron in a lower energy band it cannot send the electron up to a higher energy band and so, as the photon is not absorbed, it sails right on through. The energy of ultraviolet photons are sufficient to send electrons from the valence band to the conduction band and so they are absorbed by glass. This means you won't get a suntan when sitting in a greenhouse.

Atoms and molecules in liquids are mostly bound together as seen above so they form conduction and valence bands, and this can cause them to be transparent for the same reason.

Gases are mostly transparent because they are not in crystal form and so the energy levels of their outer electrons do not spread out into bands so very few photons are absorbed. The energy levels of atoms in gases do cause a few photons to be absorbed which causes the dark lines in the absorption spectra of light passing through gases as seen in *Chapter 10 – Atom part 2*.

Why some solids are electrical conductors

As we saw above, the energy levels of electrons in crystals smear out into bands when the atoms are so close together that their electrons shells touch each other. Also, the size of the gap between the filled and unfilled bands determines what photons can cause electrons to jump from the filled band to the unfilled band.

It is very difficult for electrons in the filled band to move through the crystal because of the congestion. Any electron in the unfilled band would find it easy to move but it takes a photon to push it there in the first place.

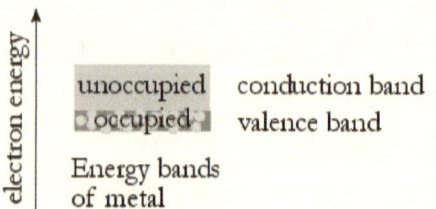

Energy bands of metal

In the case of metals, the unfilled band and the filled band overlap. This means it is easy for the outermost electrons to move through the crystal. They don't need photons to push them into the unfilled band. This is why most metals are electrical conductors.

A solid is most likely to be a good electrical conductor if it has only one electron in its outer electron shell and it readily forms crystals.

The elements that best fit these requirements are copper and silver with gold in third place. These three elements are all in one column in the periodic table.

Why electric wires get hot

With this picture of the atom and crystals we can explain why Volta's wire got hot and melted when he connected it across the terminals of his battery as mentioned in *Chapter 2 - Charge in Solids*. As electrons stream through the wire from the negative terminal towards the positive one they give a tug on the nuclei of the atoms they pass like a hand strumming the strings of a harp and, like those strings, the nuclei start to vibrate and as we saw earlier that vibration is what heat is.

In the case of a thin wire carrying a lot of electrons (high current) each atom will be 'plucked' by passing electrons more often so they will vibrate with a greater intensity. This results in the wire reaching a higher temperature. If the temperature is high enough, the peak of the black-body radiation produced will reach the visible part of the spectrum and the thin wire will start to glow. This is what happens in the filament of a light bulb or a poker that has been held in the fire too long.

Why all solids are a bit springy

The atoms in a crystal are pulled together by the attractive forces between the nuclei of one atom and the electrons of other atoms and electrons in the electron cloud of the crystal. This is very similar to the covalent bond introduced in *Chapter 11 – Chemistry*. The atoms overlap to a certain extent and their outer electrons are shared in an electron cloud (see above). But the nuclei of the atoms are all positively charged and so they repel each other. The closer they get the more strongly they repel. This means the atoms settle down in positions where the force attracting them together matches the repulsive force between their nuclei.

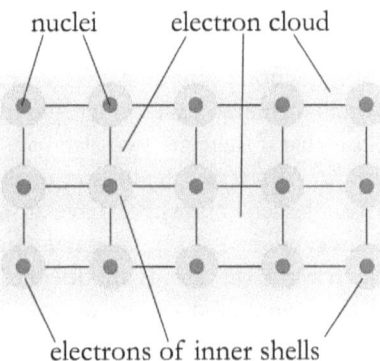

If someone places a weight on top of the crystal the force of gravity is added to the electric force pulling the atoms together. So, the atoms move closer together. However, the repulsive force between the electrons increases as the distance between them decreases. So, the atoms draw nearer until their repulsive force matches the attractive force between atoms plus the extra force of gravity provided by the weight.

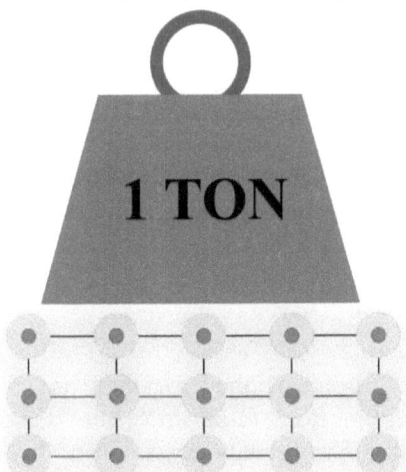

Crystals are a bit squashy because the repulsive force between nuclei increases as they get closer together. Each pair of nuclei is like a minute spring.

In this way the distance between the atoms is a measure of weight pushing down on it. And this is how your bathroom scales determine your weight.

Quantum physics explains magnetism

In *Chapter 2 - Charge in Solids* we saw how magnets were discovered and that a piece of iron can be made into a permanent magnet by rubbing a magnet against it in one direction only or by putting it in the magnetic field of a coil carrying a current.

These effects happen because electrons have a property called spin and due to the laws of induction a spinning charged particle will have a magnetic field.

Not all elements have noticeable magnetic properties and one of the main reasons for this is that if there are two electrons in an orbital they will spin in opposite directions (*Chapter 10 - Atom part 2*) and so cancel out each other's magnetic fields.

Some elements, however, have atoms with only one electron in their outer shells so these are candidates to be magnetic materials. But there are other factors which affect this, such as the way atoms are oriented within crystals and the way thermal fluctuations affect the orientation of the atoms. This leaves a few elements that exhibit strong magnetic properties, and these are iron, cobalt and nickel. Iron has two electrons in its outer shell so you would expect them to cancel out their magnetic properties but in a crystal one of these electrons will have left the atom to join the electron cloud that permeates the crystal.

These elements are known as ferromagnetic from the Latin word ferrum meaning iron.

When a piece of iron is stroked with a magnet the tiny magnetic fields of the individual atoms line up and stay lined up after the stroking has finished leaving a permanent magnet.

In this chapter we have seen how the structure of the main types of matter were discovered and how this is determined by the properties of atoms. This knowledge provided engineers with powerful tools to enrich our lives and confirmation that fundamental laws of physics are correct.

The shape of molecules

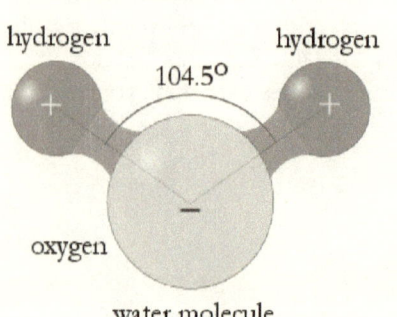

water molecule

When atoms bond together to form molecules they don't form simple lines of atoms or a ball of collected atoms. They have a particular shape. Experiments have found that water molecules are shaped like a letter 'V' with an angle of 104.5° between the two arms. This shape affects many properties of water such as the temperature at which it boils. It is also responsible for an unusual feature of water: solid water (ice) floats on water. The reason for this will be given below in the section that deals with why we have solid, liquid and gas phases of matter. The solid form of nearly all other compounds sinks to the bottom of the liquid. If ice did not float on water, the sea would freeze from the bottom up. As well as there being no place for penguins, no oil fields would have formed, and the climate would be very different.

A further aspect of the water molecule is that, because of the way electrons are distributed in the molecule, the oxygen atom has a negative charge and the hydrogen atoms have a positive charge. It is this feature of water molecules that allow them to pull atoms apart which is what happens when salt dissolves in water. This ability to dissolve substances that makes it one of the key ingredients for life to form.

A salt molecule is formed of a sodium atom (Na) and a chlorine atom (Cl) joined by an ionic bond (*Chapter 11- Chemistry*). When this bond formed the chlorine atom stole an electron from the sodium atom. This causes the sodium atom to have positive charge and the chlorine atom to have negative charge.

When a salt molecule finds itself surrounded by water molecules the oxygen in the water molecules is attracted to the sodium and the hydrogen to the chlorine. Because of the shape of the water molecule hydrogen atoms end up in a place where their proximity causes the sodium and chlorine atoms to break their ionic bond and split apart.

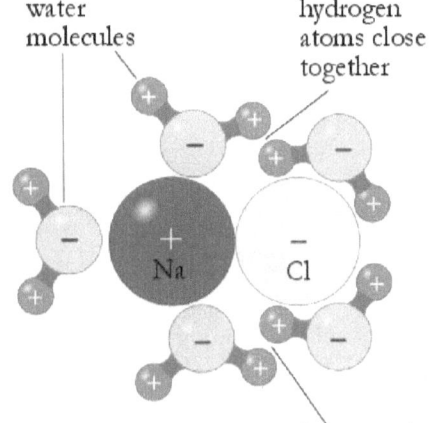

Does the sodium atom get its electron back? Oh no, greedy old chlorine keeps hold of it and goes laughing off into the sunset with its new water molecule buddies.

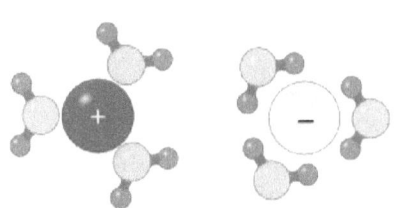

And this is how the ions that Michael Faraday had discovered were formed (*Chapter 8 - Charge in Fluids*). For the chlorine atoms the sunset was the positive electrode in one of Faraday's experiments where it caused fizzing as the chlorine gas collected into bubbles and floated up next to the electrode.

In biology the shape of molecules is much more complex and is one of the main factors which allows life to exist. It is also key to the development of drugs and the understanding of diseases.

How the water molecule got its shape

Going back to the water molecule - in *Chapter 11 - Chemistry* it was mentioned that the oxygen atom has two electron shells with two electrons in its first shell and six electrons in its outer shell.

The shape of atoms and molecules are determined by the orbitals of the electrons (*Chapter 10 – Atom part 2*). The first shell of any atom has one orbital and the second shell has four orbitals. Diagrams of electron shells are usually drawn with locations of electrons in groups of two. This is to indicate the orbitals which can accommodate two electrons each.

The orbital in shell 1 and the first orbital in shell 2 are both spherical. However, each of the remaining three orbitals of the second shell are shaped like an hourglass as in Fig 13-6.

electron shells of an oxygen atom

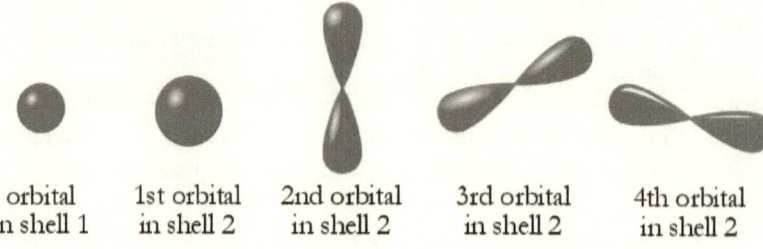

Fig 13-6 Shapes of orbitals in the first two shells of an atom

Once Schrödinger's orbitals had been discovered, (Chapter 10 – Atom part 2) it became clear that the idea of electrons being located in shells needed a little modification. Atoms with more than four electrons start to use the hourglass shaped orbital and take on a distinctly non-spherical shape (Fig 13-6).

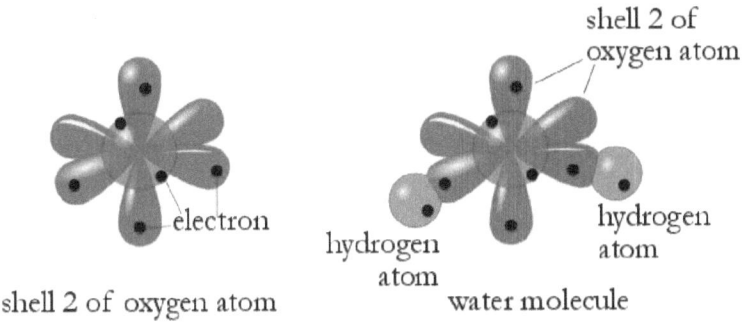

Fig 13-7 Formation of a water molecule

When hydrogen atoms bond with oxygen atoms to form a water molecule they bond to the two orbitals that only have one electron. This, coupled with the electrical repulsion between the two positively charged hydrogen atoms, gives the water molecule its 104.5° 'V' shape.

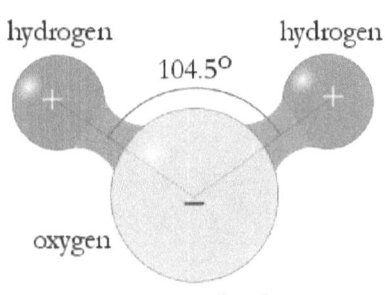

Why we have solids, liquids and gases

If you take a lump of ice and heat it, it will turn into water and if you heat the water that will turn into steam. This is true for all substances: they have a solid phase, a liquid phase and a gas phase (apart from helium which has no solid phase). These phases all have their own properties in relation to any vessel that contains them.

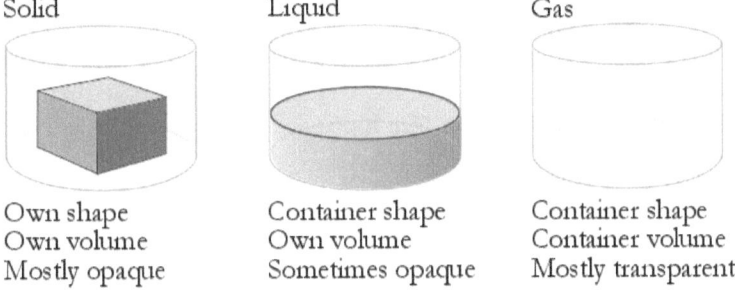

Fig 13-8 Shape, volume and transparency of solids, liquids and gases

Why is this? It is a consequence of the attractive forces between molecules.

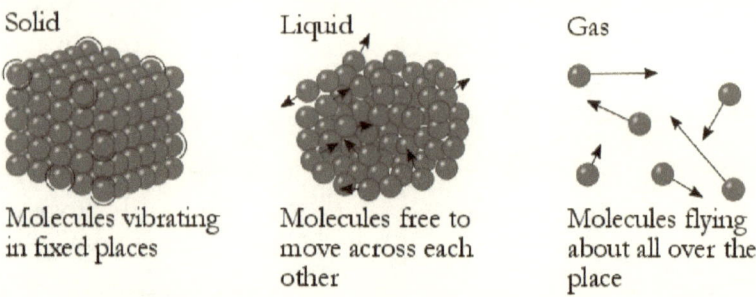

Fig 13-9 Arrangements of molecules in solids, liquids and gases

The reason these three phases of matter exist is due to a trial of strength between heat and the bond energy (*Chapter 11 – Chemistry*) between molecules[19] which is caused by the uneven distribution of charge.

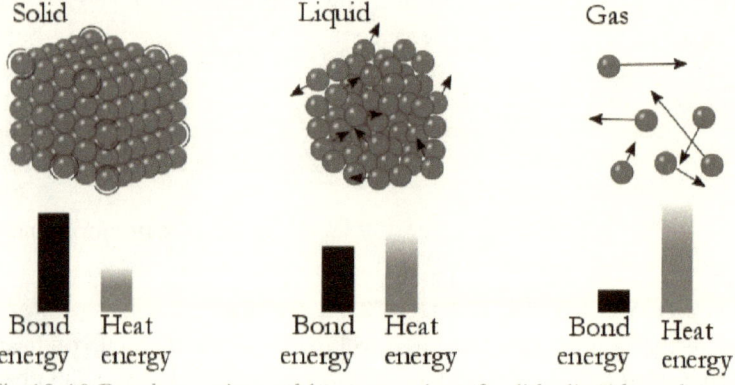

Fig 13-10 Bond energies and heat energies of solids, liquids and gases

When a substance is very cold its molecules have little kinetic energy and its molecules stay in the same place in contact with the same neighbours. They only have enough kinetic energy to dance around in the same place. The substance exists as a solid. In this situation the bond energy between molecules is greater than their kinetic energy as can be seen by the black and red bars representing bond energy and heat energy in Fig 13-10. The bond energy wins and the forces between molecules is strong enough to stop the force of gravity from pulling the substance out of shape.

When a substance is very hot its molecules have a lot of kinetic energy and its molecules fly about all over the place and never get to know their neighbours. The substance exists as a gas. In this situation the kinetic energy of the molecules is greater

---

[19] The most common of these is Van der Waals force. Other types of attractive electric force between molecules also exist.

than the bond energy between them. The kinetic energy wins and the forces between molecules is not enough to stop them flying about anywhere in the container.

When a substance is between these extremes of temperature, its molecules remain in close contact with each other but are free to move to new locations. The substance exists as a liquid. There are bonds between its molecules, but those bonds are constantly breaking and reforming. The kinetic energy is sufficient to break bonds between molecules, but collisions can cause them to lose energy and so the bonds reform. The kinetic and bond energies have an uneasy truce. The force of gravity on the molecules is enough to pull them out of shape so they fill the bottom of the container.

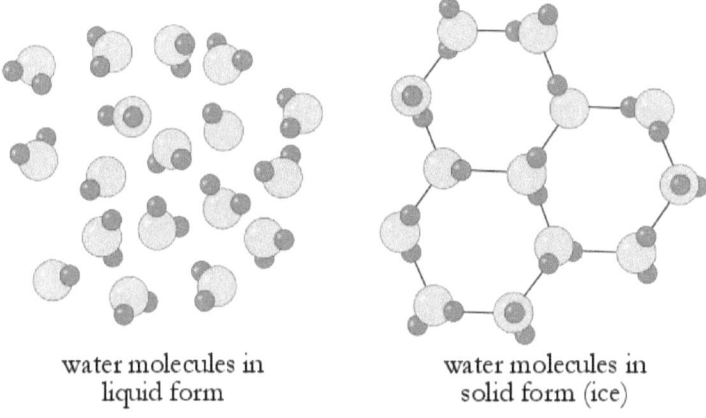

water molecules in
liquid form

water molecules in
solid form (ice)

Fig 13-11 Liquid and solid forms of water

The solid form of most substances is more dense than liquid form but there are exceptions and one of these is water. An ice cube in a glass of water will float to the top because ice is less dense than water.

So why is this? Crystals can be formed of atoms or molecules. When a crystal is formed from molecules the shape of the molecules affects the type of lattice structure of the crystal. The V shape of water molecules causes the crystal lattice to form with a hexagonal structure with a void inside each hexagon (Fig 13-11). This is what makes ice less dense than water and is the reason that ice floats on water.

A fourth phase of matter

In 1879 when William Crookes was investigating light produced by the tube that takes his name (*Chapter 8 - Charge in Fluids*) he realised that, under some circumstances, this was caused by glowing gases which he called radiant matter. Subsequent investigation showed the way that light is produced is the same way that sparks, lightning and some parts of flames produce light. We now know this as plasma (not to be confused with blood plasma).

As matter gets hotter it moves about faster and bonds between molecules break so that solids become liquids and liquids become gases. But also, the bonds between atoms can break causing compounds to separate out into their constituent elements[20]. As matter gets hotter still the bonds between electrons and the nucleus in atoms break to form ions.

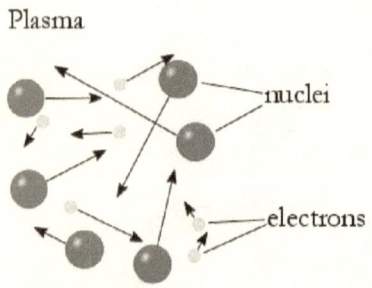

When a large proportion of electrons have left their atoms, the substance can be affected by magnetic and electric fields and, as well as glowing with black-body radiation due to the changing motion of charged particles, photons are emitted as electrons fall back into electrons shells vacated earlier. This causes photons to be emitted in the way described in *Chapter 10 - Atom part 2*. This is the plasma that Crookes saw in his tube and it is now considered a fourth state of matter after solid, liquid and gas.

Air is a poor conductor of electricity, but a plasma is a very good conductor of electricity because the electrons are free to go wherever an electric field pulls them. So, sparks and lightning allow an easy path for charge to flow through air.

In this chapter we have seen how the outer shells of atoms are responsible for nearly all the properties of matter.

---

[20] This was often used by people searching for new elements

Key points of this chapter

- In solids the heat energy of molecules is less than the bond energy holding them together, so they stay bonded in place
- In gases the heat energy of molecules is greater than the bond energy holding them together, so they do not bond but fly about all over the available space
- In liquids the heat energy of molecules is sometimes greater and sometimes less than the bond energy holding them together, so they stay in contact by sliding over each other
- If a gas gets so hot that most of the electrons leave their atoms this gas enters the fourth phase of matter which is a plasma
- Sparks and lightning are common forms of plasma
- Crystals comprise a regular lattice of molecules or atoms
- The shape of the cells of the lattice and the overall shape of the crystal are determined by the relative sizes of atoms or molecules and other factors
- The outer electrons of atoms in a crystal are shared across the entire lattice, making them good electrical conductors
- The structure of crystals is examined via X-ray diffraction
- When atoms and molecules come close together in a crystal the top two energy levels of their electrons smear out into bands which happens due to Pauli's Exclusion Principle
- The highest energy band which contains electrons is known as the valence band
- The energy band above the valence band which does not contain electrons is known as the conduction band
- Some solids are transparent to visible light because the energy in photons of visible light is insufficient to send them from the valence band to the conduction band
- As electrons in an electric current pass the nuclei of the atoms of the conductor, they cause them to vibrate and this is why the conductor heats up

14 COSMOLOGY PART 2

By studying light from objects seen in the night sky, astronomers made great progress in our knowledge of how the universe works. In this chapter we see how discoveries in other fields of physics were used by astronomers to establish the history of the universe, to understand how elements were created and to uncover new dark mysteries.

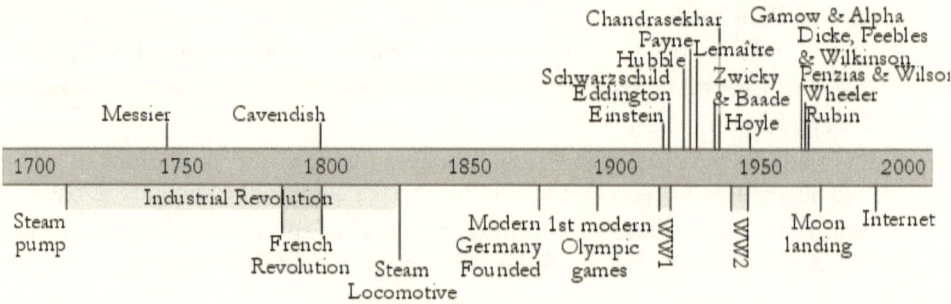

How the mass of planets was found

In *Chapter 4 - Cosmology part 1* we saw how astronomers found the distances to planets and stars. But how do we know the masses of planets? So far this book has avoided confronting the difference between the terms mass and weight by always using the term mass. But now we need to address that. The term weight is used for the force that pulls an object towards the earth and it depends on how much material is in the earth. If you take that object to the moon which comprises much less material the force of gravity would be less and so it would weigh less. Mass, on the other hand is a measure of the amount of material in an object and is the same no matter where in the universe it is. Both weight and mass are measured in kg.

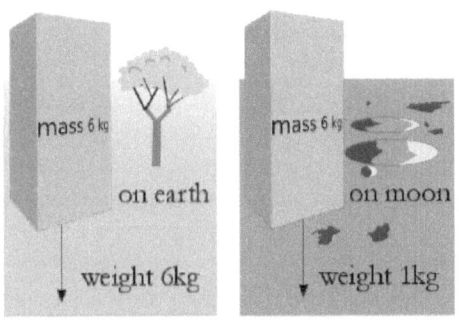

In *Chapter 1 – Energy* we saw how Newton produced an equation to determine the gravitational force between two objects if you knew the masses of those objects and a number called the gravitational constant, but he could not find the value of the gravitational constant. Without this the mass of the earth could not be established.

In 1798 British scientist Henry Cavendish found a way to determine the gravitational constant by measuring the tiny force of gravitational attraction between two lead balls[21]. This force was 50 million times less than their weight.

In doing so he demonstrated, for the first time, that gravitational force acts between two objects in a laboratory and verified Newton's law of gravity. From this experiment Cavendish was able to determine the gravitational constant which made it possible to establish the mass of the earth for the first time. Using this information and the mathematical laws of motion and planetary motion produced by Newton and Kepler (*Chapter 1 - Energy*) it was

---

[21] The actual experiment was more complex than suggested here.

possible to calculate the masses of all of the planets in the solar system. The earth weighs 600 thousand million million tonnes.

Cavendish had done for planetary masses what Millikan did for the mass of the electron (*Chapter 10 - Atom part 2*).

How galaxies and their distances were found

The story of how galaxies were discovered starts in the 1740s with French astronomer Charles Messier who was searching the sky for comets. The method he used was to find a fuzzy blob and record its position and then find it on a later night and see if it had moved. Many of the fuzzy objects he found turned out to be a distraction because although they were fuzzy he knew they were not comets because they were in the same place in the sky on subsequent nights. As a service to other comet hunters around the world he produced a catalogue of these fuzzy objects so others could save time by ignoring them. These objects are still listed in maps of the night sky as M followed by a catalogue number. Some of these were known as galaxies from the Greek word meaning milky. Messier object 31 (M31) became known as the Andromeda galaxy because it is situated in the Andromeda constellation.

In 1915 Albert Einstein produced general relativity equations that again changed our concepts of time and space. The subject of general relativity is very complex but here, all we need to concern ourselves with is that the equations predicted that the universe should be constantly expanding. This ran counter to the accepted view of the time that although planets orbit around stars the distance between stars is unchanging, a concept known as the static universe. It led Einstein to believe there was a fault in his equations and so he added a number known as the cosmological constant to his general relativity equations to make them agree with the idea of a static universe.

Later, Russian physicist Alexander Friedmann used the original general relativity equations without the cosmological constant in 1922 to build a model to explore what an expanding universe would look like. At the time little importance was placed on this work because of the general belief in the static universe.

Our idea of what the universe was like changed profoundly in 1923 when American astronomer Edwin Hubble got access to the largest telescope in the world and pointed it at the Andromeda galaxy. Hubble saw that it contained stars arranged in a spiral. One of them he found to be a Cepheid variable and using its period he calculated its distance (*Chapter 4 - Cosmology part 1*). It turned out to be the most distant object yet measured, putting it far outside our galaxy. It became apparent that our sun is in a galaxy of many stars and the Andromeda galaxy is a separate organised group of many stars and the term 'galaxy' took on a new meaning. We now know that galaxies contain hundreds of million stars.

If people had a hard time accepting how far away stars in the Milky Way galaxy were (hundreds of light years) now a further shock was in store when they found out the distance to other galaxies. Andromeda is more than 2 million light years away. We now know that it is one of our galaxy's closest neighbours. If you scaled galaxies so that the Milky Way was in London and Andromeda was in Brighton (50 miles or 80km away) they would be as thick as a piece of paper. So, there's an unimaginable distance between galaxies. At least they are fixed in space. Well that's what people thought.

Priest points to the start of the universe and its age

In the following years many more galaxies were discovered at greater distances. Hubble also contributed to the task of finding new galaxies and crucially, he also set to work measuring their distances and the speed of movement towards or away from us.

So how did he manage this? We have seen how the distance to galaxies can be measured but how about their speeds? In *Chapter 3 - Light part 1* we saw how spectral lines are used to identify gases in the outer layers of stars. This relies on the fact that each gas has a unique pattern of spectral lines. When Hubble examined the spectral lines in light coming from other galaxies, he saw the familiar patterns of hydrogen and helium lines etc. but he also saw they were not in the same place as those produced in the laboratory.

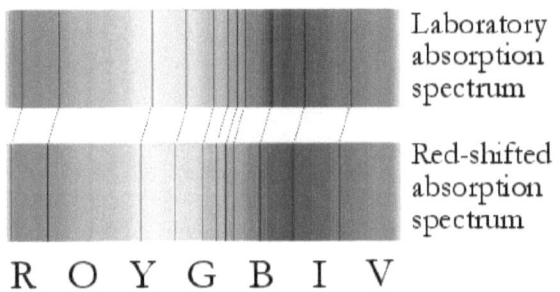

Laboratory absorption spectrum

Red-shifted absorption spectrum

R   O   Y   G   B   I   V

Some were shifted towards the blue end of the spectrum and some to the red end. Hubble realised this was due to something called the Doppler effect which causes a change in wavelength of waveforms being emitted by moving objects. The terms redshift and blueshift were introduced for light that is affected this way.

You experience the Doppler effect when you hear a motorbike pass by. As it comes towards you the pitch of its noise is high then once it has passed and is going away from you its noise changes to a lower pitch - eeeeeEEEEEEEOOOWWWWWWwwwwwww. This is because the waves emitted by the bike coming towards you get squashed up and when it is going away, they get stretched out.

So why should the waves get squashed and stretched? Each time an object emits a sound wave that wave will form a circle that spreads out from the object. If the object is stationary its sound waves will form concentric circles spreading out from the object's location. The smallest circle is for the wave that has most recently been emitted and the outer soundwave is the oldest one.

If the object moves, the sound waves continue to spread out from the location the object was when it emitted that wave. Young waves spread out from a different place from old waves. This distorts the pattern of waves emanating from the object as shown. It means the distance between waves is less for those parts of the waves in front of the object and the distance between them is greater for

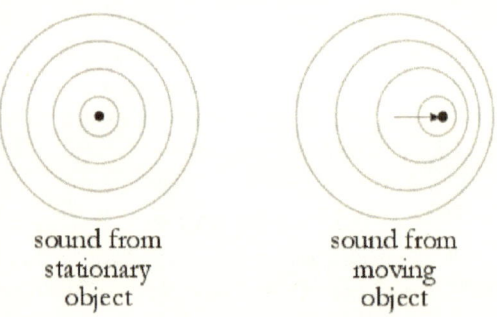

sound from stationary object

sound from moving object

those parts of the waves behind the object. So, the wavelength of the sound is less in front of the motorbike than it is behind it. Our ears detect this difference in wavelength as a difference in the pitch of the sound.

This is also true for light spreading out from a moving galaxy. The wavelength of light being sent out in the direction of the galaxy's motion will have a shorter wavelength (blueshift) and the wavelength of light emitted 'behind' the galaxy will have a longer wavelength (redshift). When Hubble examined the spectral lines in the light coming from the Andromeda galaxy he found the spectral lines are blueshifted by an amount which means it is hurtling towards us at millions of miles an hour. Luckily, it is such a long way away that we've got 3.75 million

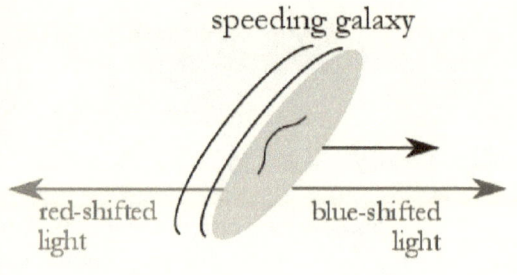

speeding galaxy

red-shifted light

blue-shifted light

years before it hits us. Hubble turned his attention to other galaxies and found the more distant ones are heading away from us. Also, he found the speed with which galaxies are receding is greater the further they are away. Gradually a map of the universe formed with clumps of galaxies moving towards or around each other and the distances between these clumps increasing.

When Belgian priest and astronomer Georges Lemaître learnt of Hubble's discovery he realised that the way galaxies moved matched the predictions of Alexander Friedmann's model of an expanding universe. Lemaître went on to explore what the expanding universe tells us about its history. He imagined a universe with time running backwards. In this case galaxies would be moving towards each other and he realised they were once all compressed into the same tiny spot, a spot he called the primeval atom. Then, running time forward again he theorised that all matter in the universe had at some time in the past exploded out of his primeval atom. However, most of the scientific community disliked this idea and found it very difficult to accept and gave it the deliberately derogatory term "Big Bang".

But as people realised how well Friedmann's model fitted Hubble's discoveries the Big Bang theory gradually became accepted by the scientific community. In 1931 Einstein accepted Lemaître's conclusions and removed the cosmological constant from the general relativity equations. More recent measurements of the speed of expansion and the distance between galaxies have established that the Big Bang occurred 13.7 billion years ago.

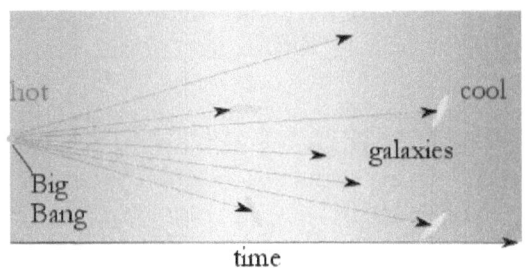

Invoking the combined gas laws (*Chapter 1 - Energy*), working backwards in time from where we are now would be like compressing a gas so it would get hotter. So, the very early universe would have been very hot. Going forwards in time from the Big Bang the universe would expand and cool.

Consequences of this hot early universe were described by American scientists George Gamow and Ralf Alpha in a paper published in 1948. It contained calculations describing what the universe would have been like in the time just after the Big Bang. Because of the extreme temperature none of the planets, stars or galaxies we see today would have existed, nor would the atoms that planets and stars are made of. This is because the protons, neutrons and electrons would have been flying about independently. They would have more kinetic energy than the attractive force between protons and electrons could overcome so the protons (which are in effect hydrogen nuclei) could not capture electrons to form atoms.

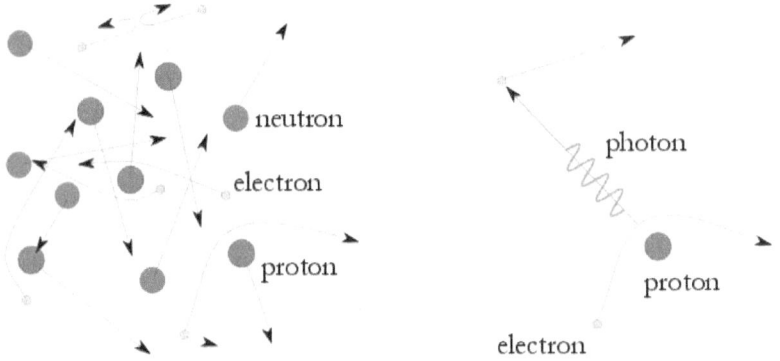

Fig 14-1 Activity of particles in the first 15 minutes of the universe

The activity of these particles in the early universe is shown in Fig 14-1. As the electrons flew about, they encountered protons and other electrons. In many circumstances this will have caused them to decelerate. Each time this happened a photon will have been emitted by the electrons as predicted by Maxwell's equations (*Chapter 3 - Light part 1*). The photons emitted this way would have filled the universe, but they would not get very far because as soon as they encountered an electron they would be absorbed which in effect passed the energy of the photon on to the recipient electron as shown in Fig. 14-2. The light produced was scattered rather like the way a dark raincloud scatters light.

At this time, although the universe was generating a lot of light, it was opaque.

Just as a gas cools as it expands, the universe would have cooled as it expanded. At about fifteen minutes after the big bang the universe would have cooled enough for protons and neutrons to stick together when they hit. This nuclear fusion (*Chapter 12 - Nucleus*) would cause some protons and neutrons to combine to form helium nuclei[22]. Gamow and Alpha's paper of 1948 gave a ratio for these two elements. This was to become an important factor in establishing how stars shine as will be seen below.

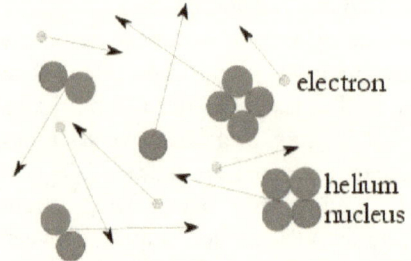

The electrons at this stage would not have joined the hydrogen nuclei and helium nuclei because the universe was still too hot, which means they were flying about too fast to be held by the attractive force of the protons.

These free flying electrons would still be decelerated as they encountered helium nuclei, protons and other electrons and so would still generate photons. And these photons would still be absorbed by free electrons. And the universe remained opaque.

At 380,000 years the universe would have cooled enough for electrons to join the hydrogen and helium nuclei and form electron shells around the first atoms. In *Chapter 10 - Atom part 2* we saw that electrons in electron shells and orbitals of atoms will not absorb just any old photon, they will only absorb photons whose energy exactly matches the difference in energy between an electron's current energy level (one energy level per electron shell) and a higher one. All other photons just sail on by. Also, now that the electrons had been captured by hydrogen and helium nuclei to form atoms there were no more free electrons to fly about and generate photons.

At 380,000 years after the Big Bang the universe became transparent and stopped producing light.

The last photons emitted before the universe became transparent would have escaped the clutches of the atoms. This would have been an unimaginably bright flash.

---

[22] Isotopes of hydrogen would also be created at this time. E.g. one proton and one neutron. Some of these would combine to form helium nuclei.

This light is known as the "last scattering".

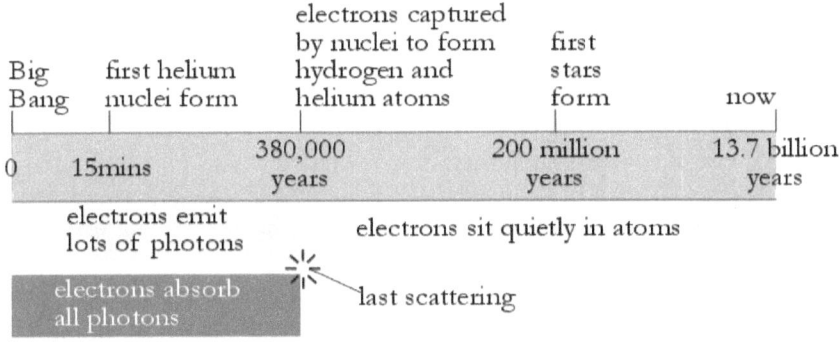

Fig 14-2 Timeline of the universe showing the last scattering

Gamow and Alpha's theory remained just that, a theory, until 1964. It was then that two American engineers Arno Penzias and Robert Wilson working for Bell Laboratories were trying to measure microwave radiation that had been bounced off two early satellites (named echo 1 and echo 2) using a very large, sensitive aerial. This returning signal was extremely weak and was drowned out by other nearby sources of microwave radiation. They successfully managed to identify these sources and shield their aerial from them one by one until there was only one left. This was a great nuisance; they could not locate it because it seemed to be coming from everywhere. They made measurements at many microwave wavelengths and produced a graph whose shape matched the black-body radiation curve (*Chapter 3 - Light part 1*).

They further calculated this is the radiation that an object 2.7° above absolute zero would emit. The problem was they saw this radiation was coming from every direction so whatever this object was it would have to be very big.

At this time American astrophysicists Robert Dicke, Jim Peebles and David Wilkinson had calculated that the light of the "last scattering" should still be visible today, 13.7 billion years after it was emitted. Using general relativity they calculated that its spectrum should by now be a black-body curve with its peak in the microwave part of the electromagnetic spectrum, just like the radiation a black-body with a temperature of 2.7° above absolute zero would emit. While they were making preparations to start their search they happened on the news that the radiation they were going to look for had already been discovered by Penzias and Wilson. It became clear that they had already accidently discovered the "last scattering" also known as the Cosmic Microwave Background.

Well done Penzias and Wilson. Have a Nobel prize.

The first astronomers only had access to the visible light part of the spectrum, but

Penzias and Wilson had unwittingly become astronomers using the microwave part of the electromagnetic spectrum. These days all parts of the electromagnetic spectrum are used by astronomers:
- Radio telescopes show to stars when they die
- Infrared telescopes show where dust is in galaxies
- Ultraviolet light shows hot young stars
- X-rays show what happens to stars when they die

The light which first spread across the universe was due to the acceleration of charged particles and it was not until the universe was 200 million years old that the first stars formed. This was when hydrogen and helium gas permeating the universe clumped together and gravity dragged in more of these gases until the conditions for nuclear fusion were met causing the first starlight to spread into the universe.

How our elements are created in stardust
So, what goes on in stars when they create nuclei?

As astronomers began to realise how distant stars were it became clear if we can see them in our sky, they must be very bright. Also, it became clear they are very old, so they have been burning bright for a long time. This means stars must contain an enormous store of energy to allow them to radiate so much energy in the form of electromagnetic radiation for such a long time. But, people wondered, what is that source of energy? All estimates based on known chemical reactions came up with figures far short of the lifetimes of stars (more than 1 billion years).

A possible solution to this mystery came in 1920 when British astronomer Arthur Eddington suggested pairs of hydrogen atoms combine to form helium in stars via nuclear fusion reactions (*Chapter 12 - Nucleus*). However, this was not accepted by the scientific community because it was then believed that there was very little hydrogen and helium in stars. Iron was assumed to be the most abundant element in stars.

And so things remained until British astronomer Cecilia Payne working in Massachusetts, USA in 1925 used astronomical data to determine the most common element in stars. She found hydrogen was the most common at 74% followed by helium at 24% with all the other elements making up the final 2%. Payne's result was dismissed until Gamow and Alpha's paper (see above) describing the Big Bang was produced which mathematically proved that just this ratio is precisely what should be expected.

Following this, in 1934, Swiss and German astronomers Fritz Zwicky and Walter Baade worked out what happens in stars as they burn their nuclear fuel. The fusion process generates heat which causes the star to expand while at the same time the gravitational attraction between the particles of the gases in the star push in and constrain the size of the star. So we have one force pushing in and the other pushing out which balance and give the star a constant size.

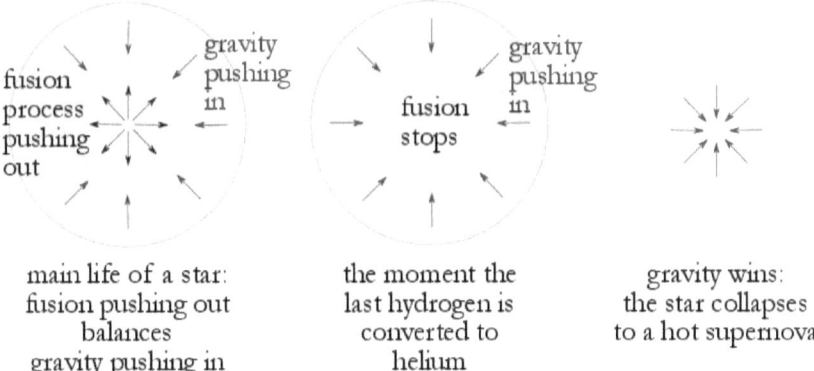

Fig 14-3 Life and death of a star

Eventually however, there comes a time when all the hydrogen in the star has been converted into helium. Now there is no more fusion reaction, no more heat being generated and no more force pushing out. But the force of gravity is still pushing in. In 1935 Indian physicist Subrahmanyan Chandrasekhar produced a paper showing that when this happens to a star with more mass than 1.5 times the mass of the sun it collapses to form a supernova. In doing so it becomes very hot again, in the same way that a compressed gas gains temperature (combined law of gases, *Chapter 1 – Energy*). So hot in fact, that it shines with the brightness of 100 million stars, as bright as an entire galaxy, but this lasts only for a few years. These violent events are the source of the cosmic rays which continually rain down on earth which among other things make carbon dating possible (*Chapter 12 – Nucleus*). The debris of supernovae is spread out to become fuzzy blobs some of which found their way alongside galaxies into Charles Messier's catalogue and some of which become new stars and planets. Then, in 1938, using data from Chinese astronomers and the Mount Wilson telescope in the USA, Baade worked out that the crab nebula which is the first entry in Charles Messier's catalogue (M1) is the remnants of the supernova seen in 1054 AD. There is also a theory that the star of Bethlehem may have been one of these supernovas.

Well done Subrahmanyan Chandrasekhar, have a Nobel prize.

The next step in understanding the life of stars came in 1946 when British astronomer Fred Hoyle produced papers which showed how the fusion process (*Chapter 12 - Nucleus*) at centres of stars can cause hydrogen nuclei to fuse to create helium[23] and how this process generates heat to allow helium nuclei to fuse and create beryllium. Hoyle's paper showed how elements from helium to iron in the periodic table (atomic

---

[23] This process also occurred after the Big Bang before stars had formed.

numbers 2 to 26) are created in the centres of stars. However, the theory did not explain how the heavier elements from cobalt to uranium (atomic numbers 27 to 92) form. This is because the fusion process to create these elements requires more heat than is provided by the reaction itself.

In 1957 Hoyle, along with 3 others produced a paper which described how the extreme conditions in supernovae can create elements heavier than iron because they have so much heat energy. If you have a ring that's made of gold, that gold was created in a supernova.

In general, the lighter the element the more abundant it is throughout the universe because heavier elements are created by fusing lighter elements together. Thus, copper is more abundant than silver which is more abundant than gold.

The fusion process is also how the energy that was released in Meitner and Hahn's nuclear fission (*Chapter 12 - Nucleus*) got into the uranium atoms in the first place. It came from the kinetic energy of the hot early universe creating helium nuclei and from stars and supernovae using their internal heat to combine those helium nuclei to nuclei of heavier elements. It is a situation where kinetic energy was turned into potential energy.

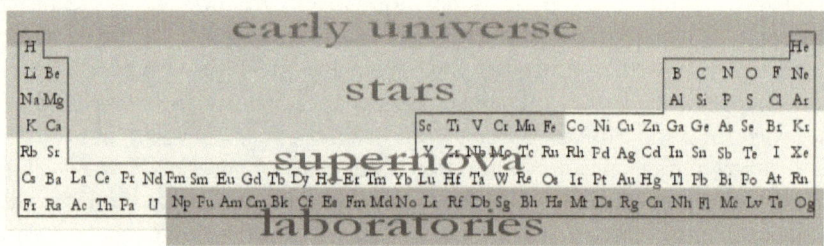

Fig 14-4 Periodic table showing where elements were created

We have seen that elements heavier than uranium are not found on earth. These are sometimes known as 'unnatural' elements or transuranic elements. There are two reasons for the lack of these elements on our planet. One is that most of these elements have a half-life that is less than a few minutes, so they change into lighter elements via radioactivity (see *Chapter 12 – Nucleus*). However, the element plutonium has an isotope with a half-life of 83 million years yet this was not found on earth until it was first created in a nuclear reactor in 1940. It is assumed this is not found on earth because the supernova which created the elements which form our planet was not hot enough to create an element this heavy.

Black holes

Since the 18th century it had been recognised that objects in space like planets and stars form because gravitational attraction of dust particles and gas cause them to clump together. The more they clump the more mass they have in one place so the stronger the attraction so the larger and more massive they become.

As dust clumps together, objects like asteroids or comets form. This process often involves several lumps coming together to form large oddly shaped rocks. As objects grow in this way they acquire more mass and so the force of gravity they exert on nearby objects increases. When such objects have grown until they have a diameter of about 300km the gravitational pull on the particles is big enough to force the object to take a spherical shape.

When hydrogen gas collects in this way the pressure of the gas at the centre increases and as it does the combined gas laws take effect (*Chapter 1 - Energy*), so the gas gets hotter. Eventually such a clump becomes hot enough for hydrogen fusion to occur. At this point energy is released and the clump of gas becomes a star.

One of the surprises provided by Einstein's 1915 general theory of relativity was that gravity attracts not only normal matter which has mass but also light. Under normal circumstances this effect is too small to be noticed but in 1919 British astronomer Arthur Eddington managed to detect and measure this effect. To do this he needed to observe a beam of light that had come close to something massive and the most massive thing in the vicinity of planet earth is the sun. Eddington calculated that light from a star that is nearly behind the sun should be bent by a measurable amount. The problem is that stars are not visible during the day so any stars nearly behind the sun are invisible.

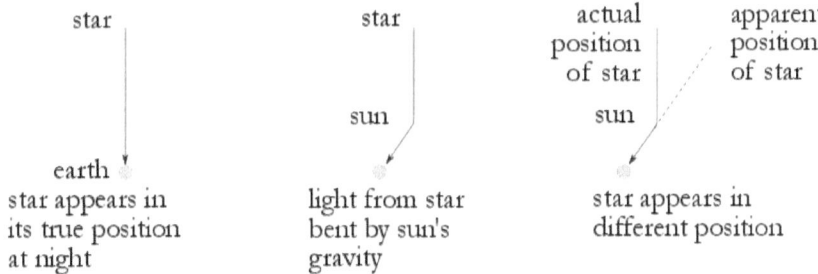

Fig 14-5 Light from star bent by sun's gravity

So Eddington set about measuring the position of a star which appears close to the sun during a solar eclipse. He could then compare its position with its position at night to see if its position had changed due to the path of its light being bent by the gravity of the sun. In 1919 he took his equipment to the island of Principe off the coast of Africa where the solar eclipse was visible. During the eclipse he measured the location of several stars that appeared in the sky near the sun and found their apparent positions differed to their night-time positions by an amount that agreed with Einstein's theory of general relativity. This was the first experimental proof that Einstein's general theory of relativity is correct.

Following work on general relativity by German physicist Karl Schwarzschild in 1916, it was realised that it should be possible for an object so massive to form that it would swallow all matter and light that came anywhere near it. Even light it emitted itself would be unable to escape its intense gravitational field. In 1965 these theoretical objects were given the name black holes by American astronomer John Wheeler.

The fact no light escapes from black holes means they cannot be observed directly in the normal way so indirect evidence must be relied on. By tracking the orbits of stars at the centre of our galaxy and applying Kepler's orbital laws, astronomers have been able to establish the existence of an object matching the characteristics of a supermassive black hole. More recently the image of the glow of gas heated by its compression as it falls into a black hole has been produced.

Earlier chapters introduced the idea that electromagnetic radiation is quantised in the form of photons. Scientists also have good theories which describe how the particles of radioactivity represent the quantisation of the energy which is associated with the weak nuclear force. The strong nuclear force also has a particle which allows it to operate and so it too has its own good quantum theory. So scientists have good quantum theories for 3 of the 4 known forces. But the fourth force, gravity, has never yielded to a theory of quantisation. It is quite probable that the conditions inside a black hole if ever they could be observed would serve up observations which could lead to such a theory.

Who ate all the mass?

At the end of the 1920s calculations based on general relativity produced a figure for how much mass there should be in the entire universe. Careful observations of a small patch of sky were extrapolated to see if the expected mass was to be seen. It wasn't. Less than 5 percent of the expected amount was seen. A disturbing mystery. More than 95% of the matter was missing.

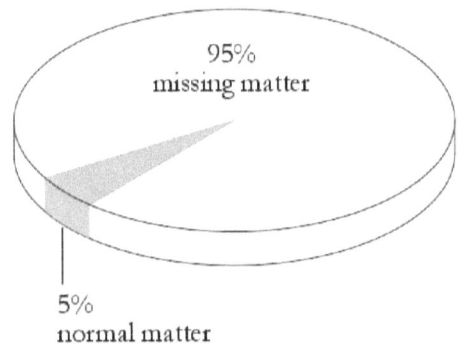

A clue to solving this mystery came in 1933 when Zwicky was studying a group of galaxies known as the Coma group. He used the redshift and blueshift of their light to determine how they were moving and found the movement did not agree with general relativity predictions. As a possible explanation for this he calculated that if a large amount of unseen mass existed around these galaxies their movement would fit the predictions of general relativity. The name dark matter was coined for this unseen and as yet undiscovered mass.

Around this time astronomers first managed to detect the redshift of stars on one side of galaxies and blueshift on the other indicating that they rotate and in the 1960s American astronomer Vera Rubin made detailed measurements of how stars' rotation speeds vary with their distance from the galactic centres. She found this speed of rotation did not agree with the predictions of Newton's laws but would agree if each galaxy had a halo of unseen matter. Her findings provided further evidence for the existence of dark matter.

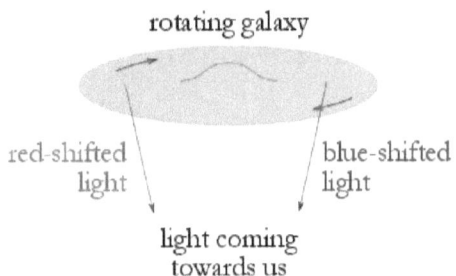

This is like the position astronomy was in after Le Vernier had calculated that the unexpected motion of Uranus pointed to an unseen planet (*Chapter 4 - Cosmology part 1*). Neptune was soon discovered but the hunt for dark matter continues today even though astronomers know where to look.

But no matter what telescopes are pointed at the regions in space where theory says dark matter should be, nothing is seen. Its substance must be quite unlike any matter we currently know. At the conclusion of *Chapter 12 - Nucleus* it was said that everything we can see is made from electrons, neutrons and protons. If dark matter exists, we can't see it and it is very unlikely to comprise these known particles. It must be a very unusual substance.

So does this account for the 95% of missing matter? No. Calculations based on the number of and sizes of galaxies in the universe indicated it accounts for 27% of the expected mass. There was still another 68% to be accounted for. Still a mystery.

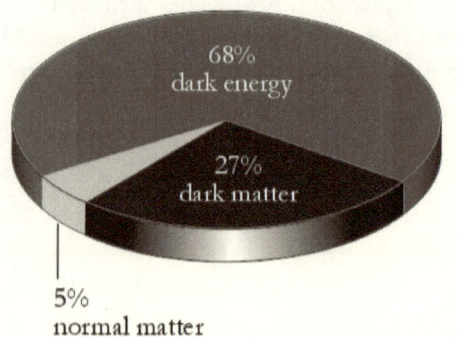

Towards the end of the 20th century two teams of astronomers who were measuring the distance to supernovae in galaxies found that the expansion of the universe was not constant. It is accelerating. For this to happen an enormous amount of energy is required. Something would be pushing groups of galaxies apart against gravity. So whatever it is, it is big. And like all energy it has some associated relativistic mass. The amount of this mass was calculated to be the same as the 68% missing mass. At last the amount of mass in the universe fits with the predictions of general relativity and the energy causing this acceleration gained the name dark energy. But dark energy is just as mysterious as dark matter. Does it imply a new force? As some questions are answered new ones present themselves.

By looking at the light coming from planets, stars and galaxies, astronomers and scientists have established the mass of planets, where elements came from and the history of the universe. But they have also detected new mysteries of dark energy and dark matter. In future, scientists and cosmologists will undoubtedly find a complete explanation for why galaxies do not move in the way general relativity predicts and, when they do, maybe that discovery will impact our lives as so many others have.

Key points of this chapter

- Apart from the closest galaxies, the distance between galaxies is constantly increasing. This was established by measuring the redshift of the light coming from distant galaxies
- The further away a galaxy is, the faster it is receding. This leads to the conclusion that all matter was compressed to a tiny point at the beginning of the universe
- The event at the beginning of the universe is known as the Big Bang
- In the first 380,000 years of the universe, electrons were moving so fast they could not be captured by protons to form hydrogen atoms. This meant the electrons did not have energy levels and could absorb light of any wavelength. This meant the universe was opaque
- After the universe was 380,000 years old it had cooled enough for electrons to be captured by protons to form hydrogen atoms. This meant the electrons had energy levels and could only absorb photons of particular energies. This meant the universe became transparent at this time. The light from that event is still visible as background microwave radiation
- Some of the hydrogen and helium was created in the early universe by nuclear fusion
- Elements up to iron in the periodic table are created in stars by nuclear fusion
- Elements from cobalt to uranium in the periodic table are created in supernova
- Elements heavier than uranium are created in laboratories
- Black holes are created when a star runs out of fuel and undergoes gravitational collapse
- Light cannot escape from a black hole so black holes can only be observed indirectly
- A supermassive black hole has been indirectly observed at the centre of our galaxy
- If we could see inside a black hole it would probably yield clues to a quantum theory of gravity
- General relativity indicates we can account for only 5% of the mass of the universe in the matter we can detect
- The way galaxies rotate indicates galaxies have haloes of matter we cannot detect. This is known as dark matter and accounts for 27% of the mass of the universe
- The rate of expansion of the universe indicates there is some unknown energy source accelerating groups of galaxies. This is known as dark energy and its mass equivalent accounts for 68% of the mass of the universe
- The total mass of normal matter, dark matter and dark energy mass equivalent add up to the amount of energy general relativity says should be in the universe so there is confidence these theories are correct

EPILOGUE

As scientists look further into the universe in distance and time and deeper into the atom, they uncover more mysteries that shroud and anticipate the next layer of knowledge.

At the end of the last chapter we saw that the way galaxies rotate, and the way groups of galaxies are moving can only be made to fit to the rest of our scientific knowledge by postulating the existence of dark mass and dark energy. It is a bit like we have found new pieces of the jigsaw puzzle and we are offering them up to the connected pieces to see if we can make them fit. There are some who think maybe there is a problem with the connected pieces, perhaps galaxies rotate the way they do because Newton's laws of motion work differently in galaxies. In the coming decades maybe we will find out.

Once these questions are answered is the work of physics complete? Just put 'unanswered physics questions' into your favourite search engine and you will find the answer is no.

In *Chapter 8 - Charge in Fluids*, *Chapter 9 - Light part 2* and *Chapter 12 - Nucleus* the electron, photon, proton and neutron were introduced. Work in the latter half of the 20th century has extended this list to include: 6 types of lepton 5 types of boson and 6 types of quark. The electron is one of the leptons, the photon is one of the bosons and the proton and neutron are both made up from 3 of the quarks. The photon carries energy to make the electromagnetic force happen. The 5th boson is the Higgs boson which give leptons and quarks their mass. This constitutes a very thin overview of what is known as the standard model of physics which is an almost perfect description of subatomic particles and how they interact. But it is not quite complete. Bosons have been identified as the energy carriers for three of the forces, electromagnetic, weak nuclear and strong nuclear. But no boson that carries energy has yet been found for the force of gravity.

We have an excellent theory covering subatomic particles and we have gravitation laws based on general relativity which encompass Newton's laws from our world out to the galactic scales and things travelling near the speed of light. But what we don't yet have is a theory which extends from general relativity to include the standard model.

When Paul Dirac applied special relativity to Schrödinger's equation in 1928 he found it predicted two types of electron, one with negative charge and one with positive charge. This anti-electron was later found and given the name positron. It was also found that when a positron meets an electron the two annihilate each other and the result is a gamma ray photon.

Since then anti-protons and anti-neutrons have been found and combined to make anti-atoms of hydrogen and helium. So, this sounds like an excellent rocket fuel, vast amounts of energy would be created by annihilating particles and their anti-particles. The problem is, if someone gave you a bottle of anti-protons they would immediately be drawn to the protons in the glass of the bottle and annihilated. The only hope is to constrain them in a magnetic field in a vacuum so they don't get near any real matter.

But there is a mystery concerning anti-matter. Current theory predicts equal amounts of matter and anti-matter should have been created at the big bang. And we should see it because it interacts with light, but we don't. Is it hiding somewhere? Or is more theoretical work needed?

And there is the question of the gravity and anti-matter. All measured attributes of anti-matter are the opposite of their ordinary matter counterparts, but it is not yet possible to collect enough anti-matter to weigh it. When this happens in the future will it have anti-gravity? That would be useful.

After Mendeleev produced the periodic table, atomic theory developed which explained the regularities the table presented. In a similar way there are patterns in the standard model and people are working on theories to explain them. One of those is string theory which postulates tiny one-dimensional strings which vibrate in different ways to produce the particles of the standard model. Currently this theory does not pass all experimental tests.

Finally, at the end of *Chapter 14 - Cosmology part 2*, the concepts of dark matter and dark energy were introduced. Theory and laboratory experiments cannot currently explain them.

These are some of the more prominent physics questions still to be answered. If you are interested there are many, many more. You could say these are pieces of the jigsaw puzzle that are at the edge of what has been done so far. They are not edge pieces.

REFERENCES

All chapters: Leeds University Physics Honours lecture notes

2-Charge in Solids
Electricity, magnetism: Annals of Science Volume 61, 2004 - Issue 3
https://en.m.wikipedia.org/wiki/Timeline_of_electrical_and_electronic_engineering
http://history-world.orga/benjamin_franklin_experiments_wi.htm
https://www.thoughtco.com/compass-and-other-magnetic-innovations-1991466

3-Light part 1
http://www-history.mcs.st-and.ac.uk/Biographies/Grimaldi.html
http://www.thestargarden.co.uk/RefractionReflectionDiffraction.html#sthash.JlF9aWpG.dpuf
http://www.college-optometrists.org/en/college/museyeum/online_exhibitions/observatory/newton.cfm
http://www.newtonproject.sussex.ac.uk/prism.php?id=15
http://www.thestargarden.co.uk/NewtonAndLight.html
http://upload.wikimedia.org/wikipedia/commons/thumb/0/0f/Interference_of_two_waves.svg/1280px-Interference_of_two_waves.svg.png
http://en.m.wikipedia.org/wiki/Infrared
http://coolcosmos.ipac.caltech.edu/cosmic_classroom/classroom_activities/herschel_bio.html
http://galileo.phys.virginia.edu/classes/252/spectra.html
http://www.chemteam.info/Electrons/Spectrum-History.html
http://www-history.mcs.st-and.ac.uk/Biographies/Kirchhoff.html

http://astro.cornell.edu/academics/courses/astro201/kirchhoff.htm
http://www.neafsolar.com/bb/spectroscopy.html

4-Cosmology part 1
Measuring the Universe, Kitty Ferguson
http://www.astronomyforbeginners.com/astronomy/howknow.php
http://www.astro.cornell.edu/academics/courses/astro201/hipparchus.htm
http://timtrott.co.uk/how-far-away-is-the-moon/
https://en.m.wikipedia.org/wiki/On_the_Sizes_and_Distances_(Aristarchus)
http://www.astronomyforbeginners.com/astronomy/howknow.php
https://historicengland.org.uk/listing/

5-Atom part 1
http://www.chemteam.info/AtomicStructure/AtNum-AtWtThread.html

7-Luminescence
http://geology.com/articles/fluorescent-minerals/
https://en.m.wikipedia.org/wiki/Fluorescence
http://www.blacklite.com/Technical/history_of_fluorescence.htm
https://www.fluorescence-foundation.org/lectures/madrid2010/lecture1.pdf
http://blacklite.com/Technical/history_of_fluorescence.htm

8-Charge in fluids
https://carnotcycle.wordpress.com/2017/02/01/carlisle-nicholson-and-the-discovery-of-electrolysis/
https://en.m.wikipedia.org/wiki/Crookes_tube#History
http://www.rsc.org/chemistryworld/2015/03/crookes-tube
http://www.furryelephant.com/content/radioactivity/discovery-electron-thomson/
https://www.boundless.com/chemistry/textbooks/boundless-chemistry-textbook/electrochemistry-18/electrolysis-132/electrolysis-of-sodium-chloride-529-3650/
https://www.plasma-universe.com/Electric_glow_discharge

9-Light part 2
http://sudhanshu935.blogspot.co.uk

https://light2015blog.org/2015/11/23/einstein-1905-from-energy-quanta-to-light-quanta/

10-Atom part 2
The Physicists, CP Snow
http://www.webelements.com/silver/atom_sizes.html
http://www.softschools.com/timelines/atomic_theory_timeline/95/
http://chemwiki.ucdavis.edu/Inorganic_Chemistry/Descriptive_Chemistry/Periodic_Trends_of_Elemental_Properties/Periodic_Properties_of_the_Elements
http://www.bl.uk/learning/cult/bodies/xray/roentgen.html
https://www.auntminnieeurope.com/index.aspx
https://www.uwgb.edu/dutchs/Petrology/WhatAtomsLookLike.HTM
https://the-history-of-the-atom.wikispaces.com
https://en.wikipedia.org/wiki/Bond-dissociation_energy
http://www.vias.org/genchem/energetics_12592_05.html
http://www.kentchemistry.com/links/Kinetics/BondEnergy.htm

12-Nucleus
The Physicists, CP Snow
https://www.aip.org/history/curie/periodic.htm
https://en.wikipedia.org/wiki/J._J._Thomson
http://www.kronometric.org/article/lume/#4.0
Proceedings of the National Academy of Sciences Vol 53 Number 3 March 15, 1965
http://www2.lbl.gov/abc/wallchart/chapters/03/4.html
http://www.nuffieldfoundation.org/practical-physics/henri-becquerel-discovers-radioactivity
https://en.m.wikipedia.org/wiki/Gamma_ray#History_of_discovery
http://hyperphysics.phy-astr.gsu.edu/hbase/quantum/moseley.html
https://www.aps.org/publications/apsnews/201206/physicshistory.cfm
http://www.circlon-theory.com/HTML/EmcFallacies.html
http://skepticsplay.blogspot.co.uk/2009/06/near-speed-of-light.html
https://en.m.wikipedia.org/wiki/Mass–energy_equivalence
https://scienceblogs.com/startswithabang/2009/04/03/is-uranium-the-heaviest-natura
https://profmattstrassler.com/2013/07/11/mass-ive-source-of-confusion/
http://www.physlink.com/education/askexperts/ae121.cfm

https://www.aps.org/publications/apsnews/200705/physicshistory.cfm
http://dev.physicslab.org/Document.aspx?doctype=3&filename=AtomicNuclear_ChadwickNeutron.xml
https://www.news-medical.net/health/Radiation-Poisoning-History.aspx
https://science.nasa.gov/astrophysics/focus-areas/what-is-dark-energy

https://physics.stackexchange.com/questions/24869/what-is-dark-energy-and-how-was-it-discovered
https://www.universetoday.com/115991/when-did-the-first-stars-form/
https://en.m.wikipedia.org/wiki/Chronology_of_the_universe

13-Atom part 3
http://www.chemheritage.org/discover/online-resources/chemistry-in-history/themes/molecular-synthesis-structure-and-bonding/lewis.aspx
http://www.ptable.com/
http://www-outreach.phy.cam.ac.uk/camphy/xraydiffraction/xraydiffraction_index.htm
https://www.chemguide.co.uk/atoms/structures/molecular.html
https://ci.coastal.edu/~sgilman/ice.jpg

14-Cosmology part 2
Measuring the Universe, Kitty Ferguson
The Life of Stars, Giora Shaviv
http://www.amnh.org/education/resources/rfl/web/essaybooks/cosmic/p_payne.html
http://www-istp.gsfc.nasa.gov/stargaze/Ls7adisc.htm
http://www.astro.ucla.edu/~wright/CMB.html

GLOSSARY

Absorption lines Dark lines that exist in a spectrum because light has passed through a gas which has absorbed light of certain wavelengths.

Anode A piece of metal placed in a fluid and connected to the positive terminal of a battery. See Electrode.

Atom The smallest piece of an element that can exist.

Black-body radiation Radiation emitted by objects which are hotter than their surroundings. It has a spectrum of a particular shape which makes it easy to identify.

Bond An attractive force between two or more particles which causes them to 'stick' together.

Bond energy The amount of energy required to split two molecules or atoms apart. It is equal to the amount of energy that comes out when two particles bond together.

Brownian motion The motion of pollen-sized particles suspended in water which is a result of being jostled by trillions of molecules.

Cathode A piece of metal placed in a fluid and connected to the negative terminal of a battery. See Electrode.

Cepheid variable A type of star which regularly changes brightness and whose overall brightness is related to the frequency with which its brightness changes. Useful in the measurement of stellar distances.

Chain reaction A situation where one reaction causes the same reaction to happen in neighbouring particles which go on to cause the same reaction in other particles.

Charge A property of matter which causes an object or particle to be attracted or repelled by another object or particle which has this property.

Charge carrier A charged particle which can move through a solid or fluid.

Compound A substance which comprises molecules whose atoms are of different types.

Covalent bond Bond formed when two atoms share an electron.

Electric field An area where a small charged object feels the pull of a force due to some other electric charge. Also, a representation of the strength and direction of force a charged object would feel at any place within a given area. See Force Field.

Electricity Forces caused by charge

Electrode A piece of metal placed in a fluid and connected to one terminal of a battery.

Electrolysis Process where electricity passes through a liquid and causes a chemical reaction and material to be deposited on an electrode.

Electromagnetic radiation Radiation made of electromagnetic waves that can travel through a vacuum. The most common form of which is light.

Electromagnetic spectrum A scale of all types of known electromagnetic radiation arranged in order of wavelength from radio to gamma rays.

Electron A particle that possesses negative charge and is present in all atoms (except fully ionised atoms).

Electron shell A space around the nucleus of an atom where a number of electrons can reside. See Orbital.

Electroscope A device which detects the presence of charge.

Element A substance which cannot be changed into a different substance by chemical processes or heating and which comprises atoms of only one type.

Emission lines Bright lines that appear in a spectrum because certain wavelengths have been emitted by a hot gas.

Energy The ability to supply a force.

Energy level The amount of potential energy an electron has due to the attractive force pulling it towards the nucleus. Only certain amounts of electron energy are allowed due to the effects of quantum mechanics.

Force The ability to change the speed and or direction of a stationary or moving object.

Force field An area where a small object feels the pull of a force by wind, electric charge etc. Also, a representation of the strength and direction of force an object would feel at any place within a given area.

Friction A phenomenon which impedes the progress of one object sliding over another and which causes those objects to heat up.

Gamma rays Electromagnetic radiation whose wavelength is shorter than X-rays. This has the shortest wavelength of all known electromagnetic radiation.

Geiger counter A device which detects and measures ionising radiation

Heisenberg's uncertainty principle Principle that states there is a fundamental limit to the accuracy with which we can know the momentum and position of a particle at the same time.

Infrared Electromagnetic radiation whose wavelength is longer than visible light but shorter than microwaves.

Ion An atom which has less electrons, or more electrons than protons, thus giving it positive or negative overall charge (positive ion or negative ion).

Ionising radiation Radiation whose energy is sufficient to knock an electron out of an atom and turn it into a positive ion.

Ionic bond Bond formed when one atom steals an electron from a nearby atom causing them to become electrically charged so they attract each other.

Isotope A set of atoms belonging to one element which all have the same number of neutrons.

Kinetic energy The energy an object has because it is moving.

Law A pattern of behaviour of the universe that has been seen to be the same no matter where or when it occurs.

Leyden jar An early device for storing charge.

Luminescence Phenomenon where cold material glows with one colour while being exposed to light of a colour with a shorter wavelength. In some cases, this glow persists after the incident light has been removed.

Magnetic field An area where a small iron object feels the pull of a force due to a nearby magnet. Also, a representation of the strength and direction of force an iron object would feel at any place within a given area. See Force Field.

Magnetism Forces caused by the movement of charge.

Maxwell's equations A set of 4 equations which describe how electric and magnetic fields can change in space and time.

Microwaves Electromagnetic radiation whose wavelength is longer than infrared light but shorter than radio waves.

Molecule A particle which comprises multiple atoms which may or may not be of different types.

Newton's law of gravity A law that says all objects with mass attract each other with a force that depends on the mass of the objects and the distance between them.

Newton's laws of motion Three laws which describe how an object moves in response to a force acting on it.

Neutron A particle with no charge which occurs in the nucleus of nearly all atoms.

Orbit The path an object in space takes around a larger object. This path is one in which the centrifugal force due to its motion exactly matches the force of gravity attracting the two objects together.

Orbital A mathematical construct which defines the probability of the location of a single electron or a pair of electrons in an atom. An electron shell contains one or more orbitals.

Parallax A method to determine the distance to an object by observing it from separate places. It requires the user to know the distance between the two places and the angles between the object and a line between the two places of observation.

Periodic table A table in which elements are arranged in order of atomic mass. The columns of the table hold elements of similar properties underlining the fact that elements arranged in this way repeat properties periodically

Photon A particle of light (electromagnetic radiation).

Potential energy Energy that exists when an object experiences two opposing forces.

Power The rate at which energy is converted from one form to another.

Proton A particle with positive charge which occurs in the nuclei of all atoms.

Radiation A feature which causes energy to flow through space to distant locations. Examples.: light, radio waves, neutron radiation.

Radio waves Electromagnetic radiation whose wavelength is longer than microwaves. This has the longest wavelength of all known electromagnetic radiation.

Reaction A process where a combination of substances changes into a different set of substances. This may require heat to make it happen and it may give heat off in the process.

Resonance Every object has a frequency with which it prefers to vibrate. Resonance occurs when an object spontaneously starts to vibrate when it is exposed to vibrations of the same frequency.

Schrödinger equation An equation which allows the properties of any electron orbital to be calculated, including its shape.

Spectroscope A device which allows a spectrum of light to be examined in detail.

Spectrum The visible colours arranged in order of wavelength. See Electromagnetic Spectrum.

Subatomic Smaller than an atom.

Ultraviolet Electromagnetic radiation whose wavelength is shorter than visible light but longer than X-rays.

Young's slits An experiment in which light is shone through two slits to prove that light is a waveform.

X-rays Electromagnetic radiation whose wavelength is longer than ultraviolet light but shorter than gamma rays.

INDEX

A

absorption lines, 59, 60, 62, 126, 127, 128, 211
Al Biruni, 14, 18
Alpha, 132, 133, 162, 193, 196
alpha rays, 117, 118, 119, 132, 157, 158, 162, 169
Ampere, 42, 44, 62
anode, 99, 100, 101, 102, 105, 121, 122, 130, 155, 156, 211
Archimedes, 4
Aristotle, 4, 81, 86, 87
atom, 16, 62, 133, 144, 211
aufbau principle, 128
Avogadro, 84, 114

B

Baade, 196, 197
Becker, 157, 158, 159
Becquerel, 114, 116, 117, 118
Bernoulli, 15
Bessel, 75
beta rays, 117, 157, 161, 162, 163, 169, 171
Bethe, 169, 170
Big Bang, 20, 193, 194, 196, 197
black-body radiation, 61, 64, 108, 109, 111, 178, 186, 195, 211
Bohr, 124, 125, 126, 127, 128, 130, 132, 134, 140, 147
bond, 150, 184, 209, 211
bond energy, 150, 151, 152, 187, 211
Bothe, 157, 158, 159
Boyle, 15, 82
Bragg, 173
Brahe, 9, 70
Brand, 86
Brown, 87
Brownian Motion, 87, 92, 114, 211

C

Canton, 96
Carnot, 17
Cassini, 71
cathode, 99, 100, 101, 102, 105, 114, 121, 156, 211
Cavendish, 82, 102, 189, 190
Cepheid Variable, 76, 77, 190, 211

Chadwick, 158, 159
chain reaction, 152, 168, 211
charge, 27, 28, 31, 32, 33, 38, 40, 46, 47, 62, 63, 99, 100, 102, 107, 111, 114, 117, 119, 132, 133, 144, 146, 158, 165, 167, 186, 205, 211
charge carrier, 27, 29, 30, 31, 35, 99, 105, 211
Charles, 15, 35, 190, 197
compound, 83, 85, 143, 211
Coulomb, 35
covalent bond, 147
Crookes, 100, 101, 102, 118, 185, 186, 208
Crookes tube, 100, 101, 102, 103, 110, 121, 131, 155, 156, 157
Curie, 116, 158

D

Dalton, 84
Davisson, 134
de Broglie, 134, 136, 137
Descartes, 49, 53, 108, 110
Dicke, 195
du Fay, 25

E

Eddington, 196
Einstein, 16, 64, 92, 106, 108, 109, 110, 124, 168, 169, 190, 193, 199
electric field, 41, 47, 62, 63, 65, 114, 158, 186, 212
electricity, 22, 23, 25, 26, 28, 34, 35, 38, 39, 40, 43, 63, 64, 98, 114, 178, 186, 212
electrode, 99, 105, 212
electrolysis, 100, 101, 208, 212
electromagnetic radiation, 63, 64, 65, 107, 109, 110, 111, 118, 120, 121, 132, 212, 214
electromagnetic spectrum, 63, 106, 107, 111, 118, 177, 196, 212
electron, 100, 102, 104, 110, 114, 119, 120, 127, 128, 129, 134, 138, 144, 153, 156, 205, 212
electron shell, 138, 139, 142, 147, 178, 212, 213
electroscope, 33, 107, 212

element, 81, 82, 84, 85, 87, 89, 100, 116, 124, 128, 131, 138, 144, 153, 159, 165, 170, 212
emission lines, 60, 61, 122, 128, 212
energy, 3, 11, 12, 17, 19, 20, 21, 53, 55, 64, 107, 108, 111, 113, 120, 125, 142, 151, 152, 162, 165, 169, 212
energy level, 125, 126, 127, 128, 129, 130, 131, 138, 141, 142, 147, 212
Eratosthenes, 68
Euclid, 48

F

Fajans, 157, 159
Faraday, 40, 44, 45, 62, 84, 99, 101, 131, 156, 181
force, 3, 5, 6, 7, 8, 11, 12, 14, 21, 23, 25, 26, 28, 33, 40, 45, 47, 64, 125, 145, 150, 151, 154, 165, 166, 167, 169, 171, 189, 194, 200, 212
force field, 40, 47, 212
Franklin, 26, 28, 32, 35, 36, 38, 175
Fraunhofer, 59, 60, 89, 124, 173
Fresnel, 57
friction, 5, 14, 38, 212

G

Galileo, 70, 76
Galvani, 36, 37
gamma rays, 118, 158, 162, 169, 205, 212
Gamow, 193, 194, 195, 196
gas laws, 15, 92, 193, 199
Gay-Lussac, 15, 84
Geiger, 157
Geiger counter, 104, 157, 212
Geissler, 100
Geissler tube, 100, 104
Germer, 134
Gilbert, 24, 33, 144
Goldstein, 156, 159
Goodricke, 76
gravity, 12, 14
Gray, 23, 27, 33

Grimaldi, 48, 49, 58, 207

H

Hahn, 165, 167, 198
Hauksbee, 34
heat, 11, 15, 17, 18, 19, 20, 21, 53, 54, 55, 64, 83, 84, 86, 96, 108, 109, 126, 127, 143, 150, 151, 152, 154, 168, 169, 178, 187
Heisenberg, 138, 140, 141, 147, 149
Heisenberg's Uncertainty Principle, 140, 212
Herschel, 53, 55, 71
Hertz, 106, 107, 118
Hipparchus, 69, 71
Hoyle, 197, 198
Hubble, 190, 191, 192
Huygens, 49, 50, 63

I

infrared, 53, 54, 55, 56, 61, 63, 71, 96, 107, 108, 110, 111, 196, 212, 213
ion, 99, 104, 105, 131, 142, 144, 145, 157, 212
ionic bond, 144, 145, 181
isotope, 159, 160, 161, 162, 163, 171, 198, 213

J

Janssen, 62
Joliot-Curie, 158
Joule, 17

K

Kepler, 9, 70, 71, 86, 103, 200
kinetic energy, 12, 13, 16, 17, 20, 21, 53, 64, 118, 121, 126, 127, 131, 150, 152, 166, 168, 169, 171, 198, 213
Kircher, 95
Klaproth, 86

L

Lavoisier, 82, 84
law, 4, 5, 6, 7, 9, 15, 35, 61, 71, 77, 92, 95, 121, 168, 180, 193, 199, 200, 201, 213
Lecoq, 89
Leibniz, 3
Lemaître, 190
Leyden jar, 34, 35, 36, 38, 132, 213
Lockyer, 62
luminescence, 94, 96, 128, 213

M

Magnes, 22, 43
magnetic field, 40, 41, 42, 43, 44, 45, 46, 47, 62, 63, 65, 102, 103, 116, 117, 129, 156, 180, 205, 213
magnetism, 23, 24, 25, 26, 40, 43, 46, 63, 64, 180, 213
Maxwell, 62, 64, 106, 111, 118
Maxwell's equations, 63, 65, 107, 121, 125, 131
Maxwell's Equations, 213
Maxwell's equations, 64, 111, 120
microwaves, 107, 213
Millikan, 114, 190
molecule, 15, 16, 17, 20, 53, 64, 84, 85, 86, 92, 93, 99, 114, 131, 143, 150, 151, 152, 164, 180, 182, 187, 213
Moseley, 121, 122, 124, 130, 133
Musschenbroek, 34

N

neutron, 158, 159, 160, 161, 163, 165, 166, 167, 168, 170, 171, 194, 202, 205, 213
Newton, 4, 5, 6, 7, 8, 9, 11, 13, 21, 51, 52, 53, 64, 71, 72, 77, 108, 110, 120, 141, 189, 201, 205, 213
Newton Lewis, 144

O

orbit, 10, 71, 74, 75, 76, 77, 120, 125, 132, 200, 213
orbital, 137, 138, 140, 141, 142, 147, 174, 176, 180, 182, 183, 194, 212, 213, 214

Ø

Ørsted, 40, 42, 44, 62

P

parallax, 66, 68, 69, 71, 76, 77, 213
Pauli, 138, 140, 141, 147, 149, 187
Payne, 196
Peebles, 195
periodic table, 90, 91, 92, 93, 121, 122, 124, 138, 139, 140, 142, 148, 149, 178, 206, 213
photon, 109, 110, 118, 124, 126, 127, 128, 129, 130, 131, 132, 134, 140, 142, 176, 177, 178, 186, 214
Pigott, 76
Planck, 106, 108, 109, 110, 111, 124, 127, 137, 141
potential energy, 12, 13, 16, 20, 21, 125, 126, 127, 148, 152, 165, 170, 171, 198, 214
power, 17, 19, 20, 23, 141, 170, 214
Priestley, 82
proton, 132, 133, 134, 140, 142, 144, 147, 157, 158, 159, 160, 162, 163, 165, 167, 168, 171, 193, 205, 214

R

radiation, 53, 54, 56, 61, 63, 71, 106, 108, 110, 111, 115, 116, 117, 118, 133, 158, 162, 163, 170, 178, 186, 214
radio waves, 64, 107, 111, 214
reaction, 83, 84, 132, 133, 143, 145, 159, 160, 166, 168, 214
resonance, 214

Ritter, 55, 99
Rømer, 74
Röntgen, 110, 111, 115
Rubin, 201
Rumford, 17
Rutherford, 117, 118, 119, 121, 132, 133, 158

S

Scheele, 82
Schrödinger, 137, 140, 141, 147, 149, 153, 154, 182, 205
Schrödinger equation, 141, 214
Schwarzschild, 200
Shen Kou, 45
Soddy, 157, 159
spectroscope, 59, 60, 89, 116, 214
spectrum, 51, 53, 54, 55, 59, 60, 61, 62, 63, 65, 89, 108, 128, 178, 214
Stokes, 95, 97
Stoney, 100, 102
subatomic, 157, 159, 205, 214

T

Thales, 22, 27, 28
Thomson, 102, 103, 114, 117, 121, 134, 155, 156, 157, 159, 209

U

ultraviolet, 55, 61, 63, 64, 94, 95, 96, 97, 99, 101, 102, 104, 108, 110, 111, 177, 196, 214

V

vibration, 17
Villard, 117
Volta, 37, 38, 39, 40, 99, 145, 146, 178
von Laue, 173
von Mayer, 17

W

Wheeler, 200
Wigner, 165
Wollaston, 55, 56, 59, 124

X

X-rays, 110, 111, 115, 116, 118, 121, 130, 196

Y

Young, 56
Young's slits, 57, 58, 59, 134, 214
Yukawa, 165, 166

Z

Zwicky, 196, 201

www.ingramcontent.com/pod-product-compliance
Lightning Source LLC
Chambersburg PA
CBHW030619220526
45463CB00004B/1343